HYDROCHARITACEAE

David Simpson

Monoecious, dioecious or hermaphrodite, submerged or rarely floating, freshwater or marine, annual or perennial herbs. Roots mostly simple, adventitious, rarely with root hairs. Stems corm-like or elongate, rhizomatous, stoloniferous or erect, simple or dichotomously branched. Leaves radical, or spirally arranged, or in whorls along the stem, or distichous or rarely opposite, sessile or petiolate, sometimes sheathing at the base; venation parallel or only midrib present. Stipules rarely present. Nodal scales (squamulae intravaginales) often present, situated in leaf-axils. Flowers unisexual or bisexual, 1–many, mostly regular, arranged in a spathe, the spathe axillary, sessile or pedunculate and composed of 2 wholly or partly connate, rarely free, bracts. Perianth-tube (hypanthium) often present in ♀ and ♂ flowers, exerted from or near the apex of the ovary, usually extending to carry perianth to water surface. Perianth segments 3 or 6, the latter differentiated into sepals and petals; sepals free, usually green or whitish, often reflexed; petals free, often showy, sometimes reflexed; stamens 2–many, in 1 or more whorls; anthers 2–4-thecous, basifixed, dorsally or latrorsely, rarely introrsely or extrorsely dehiscent; filaments filiform or flattened, linear or club-shaped in outline, rarely 0; staminodes often present in ♀ flowers, rarely so in ♂ flowers, in the latter the staminodes differentiated from the inner whorl of stamens; ovary inferior, composed of 2–15 connate carpels, 1(–3)-locular; placentation parietal, rarely basal; ovules few to many, anatropus or orthotropus; styles 2–15; stigmas 2–15, entire or 2-lobed. Fruit a capsule, sometimes beaked from remains of the perianth-tube, dehiscent or opening by decay of the pericarp. Seeds usually minute; embryo straight with inconspicuous or conspicuous plumule; endosperm 0.

16 genera and over 70 species, widespread throughout temperate to tropical regions. Only 3 genera are marine, the remainder occurring in fresh or brackish water. Many species show a wide range of phenotypic variation, and can be difficult to identify. Living material should be used for identification wherever possible, since most species do not dry well.

1. Freshwater plants 2
   Marine plants 7
2. Leaves in a rosette on a very short stem 3
   Leaves spirally arranged or in whorls, rarely opposite, on
   elongated stems 5
3. Leaves petiolate, leaf-blades usually more than 3 cm. wide;
   spathes often winged, usually more than 5 mm. wide   **6. Ottelia**
   Leaves sessile, less than 1.2 cm. wide; spathes never
   winged, usually less than 5 mm. wide 4
4. Leaves attenuate towards apex, 1.5 mm. or less wide, 5 mm.
   below the apex; ♂ flowers remaining attached to plant   **4. Blyxa**
   Leaves ± abruptly contracted, 2 mm. or more wide 5 mm.
   below the apex; ♂ flowers released before anthesis   **5. Vallisneria**
5. At least some leaves spirally arranged on stem; leaves with
   2 apical spines . . . .   **1. Lagarosiphon**
   All leaves in whorls, rarely opposite; leaf with 1 apical
   spine 6
6. Leaves 2–15 mm. long, often recurved; petals minute, as
   long as or shorter than sepals . . . .   **2. Hydrilla**
   Leaves ± 15–30 mm. long, never recurved; flowers showy,
   petals 3 times as long as sepals . . . .   **3. Egeria**

1

7. Plants delicate; leaves petiolate, in pairs, mostly without
distinct sheaths . . . .   **9. Halophila**
Plants robust; leaves sessile, distichous, distinctly
sheathing 8
8. Rhizome 10 mm. or more in diameter, covered in fibrous
remains of old leaf-sheaths; roots numerous, crowded,
without root-hairs . . . .   **7. Enhalus**
Rhizome 2–5 mm. in diameter, without fibres; roots 1 per
node, with root-hairs . . . .   **8. Thalassia**

1. LAGAROSIPHON

Harvey in Hook., Journ. Bot. 4: 230 (1841); Wright in F.T.A. 7: 3 (1897); Symoens & Triest
in B.J.B.B. 53: 441 (1983)

Dioecious, submerged, freshwater, perennial herbs. Roots simple, adventitious, without root hairs, arising from nodes. Stems elongate, stoloniferous or erect, filiform to terete, simple or branched from axils; branches 1 per axil. Leaves sessile, spirally arranged, subopposite or in whorls, linear to lanceolate, shortly acuminate to subobtuse, patent to strongly recurved, firm or flaccid, with 2 apical spines, often with 2 bands of sclerenchyma fibres on either side of midrib and 2–6 rows of submarginal sclerenchyma fibres; venation reduced to midrib only; margins green or hyaline, denticulate. Stipules 0. Nodal scales 2 per leaf, linear to widely ovate, with entire or papillose margins. Flowers unisexual. Male spathe solitary in leaf-axil, sessile, ovate to suborbicular, toothed, containing up to 40 flowers. Male flowers shortly pedicellate, becoming detached and floating before anthesis; sepals and petals 3, oblong-ovate to ovate, obtuse or cucullate, usually reflexed at anthesis; stamens 3; anthers 4-thecous, fixed at right-angles to the filaments; filaments filiform, spreading horizontally; staminodes often present, 3, filiform, papillose, longer than the stamens, usually joined at the apex and functioning as a sail. Female spathe solitary in leaf-axil, sessile, narrowly ovate to ovate, toothed, 1-flowered. Female flower: sepals and petals 3, elliptic to ovate, obtuse or cucullate, usually reflexed at anthesis; staminodes 3, minute, linear; ovary 1-locular, placentation parietal; ovules 5–30, orthotropus; perianth-tube filiform, exerted laterally from the ovary near apex of the spathe, carrying perianth to the water surface; styles 3; stigmas 6, 2 per style, linear, papillose. Fruit ovate, beaked at apex, pericarp smooth, irregularly dehiscent. Seeds narrowly ellipsoid to ellipsoid, smooth, with short stipe at base, attenuate at apex.

A genus of 9 species native to tropical and southern subtropical Africa. One species, *L. major* (Ridley) Moss has been widely introduced into Europe and New Zealand. Hybrids may occur where two species grow in the same vicinity, although it is often difficult to determine the parent species from such material.

1. Leaves firm, opaque, usually more than 2 mm. wide   *4. L. ilicifolius*
Leaves flaccid, translucent, rarely more than 1.8 mm.
wide 2
2. Leaf-margins green, without sclerenchyma fibres; teeth of
leaf-margin with a distinct triangular basal portion   *1. L. cordofanus*
Leaf-margins hyaline, with 2–6 rows of sclerenchyma
fibres; teeth without a distinct basal portion 3
3. Leaves over 1.2 mm. wide; leaf-margins with 4–6 rows of
sclerenchyma fibres . . . .   *3. L. hydrilloides*
Leaves mostly 1 mm. or less wide; leaf-margins with 2–3(–4)
rows of sclerenchyma cells . . . .   *2. L. muscoides*

1. L. cordofanus *Caspary* in Pringsh. Jahrb. Wissensch. Bot. 1: 504 (1855); T. Durand & Schinz in Consp. Fl. Afr. 5: 1 (1895); Wright in F.T.A. 7: 4 (1897); Symoens & Triest in B.J.B.B. 53: 456, fig. 5 (1983). Type: Sudan, 'Arasch-Cool', *Kotschy* 170 (B, holo.†, BM!, G, K!, L!, MPU, P, iso.)

Stems filiform or terete, 0.2–2 mm. in diameter, whitish to mid-green. Leaves spirally arranged to subopposite or in whorls, linear to narrowly linear-lanceolate, shortly

FIG. 1. *LAGAROSIPHON CORDOFANUS* — 1, habit, × ⅔; 2, male flower, × 12; 3, female flower, × 12; 4, fruits, × 6; 5, seeds, × 20; 1,3,4, from *Polhill & Paulo* 846; 2, after Symoens & Triest (1982); 5, from *Greenway & Polhill* 11641. Drawn by Christine Grey-Wilson.

to narrowly acute, light to dark green, patent, spreading or somewhat recurved, flaccid, translucent, 4–29 mm. long, 0.25–1(–1.8) mm. wide, with 2 broad bands of sclerenchyma fibres along either side of midrib; submarginal fibres absent; margins green, denticulate, the teeth usually curved towards the leaf-apex and with a triangular green base. Nodal scales ovate to broadly ovate, the margins entire. Male spathe ovate to broadly ovate, 1.3–2.7 mm. long, 1–2 mm. wide, pinkish white, containing 7–14 flowers. Male flowers: sepals and petals elliptic, ± 1 mm. long, 0.3–0.5 mm. wide, white, translucent; stamens ± 2 mm. long. Female spathe narrowly ovate to ovate, 1.7–2.8 mm. long, 0.7–1.2 mm. wide. Female flower: sepals and petals elliptic to obovate, 0.7–1.2 mm. long, 0.5–0.8 mm. wide, white, translucent; ovary ± 2 mm. long, with 10–40 ovules; perianth-tube up to 50 mm. long; stigmas 1.3–1.7 mm. long, purplish. Fruit ovoid, 2.5–4 mm. long, 1.5–2 mm. wide. Seeds narrowly ellipsoid, 1–1.5 mm. long, 0.3–0.5 mm. wide, brown. Fig. 1.

UGANDA. Teso District: Serere, July 1924, *Synge* 744!
KENYA. Masai District: Masai Mara, 24 Sept. 1947, *Bally* 5456!; Teita District: Mudanda Rock, 1 June 1970, *Faden, Glover & Evans* 70/185!; Kilifi District: between Mariakebuni and Marafa, 23 Nov. 1961, *Polhill & Paulo* 846!
TANZANIA. Iringa District: Kinyantupa, 2 May 1970, *Greenway & Kanuri* 14442! & Trekimboga track, 6 May 1970, *Greenway & Kanuri* 14459!; Lindi District: Nachingwea, Namatumbuzi rocks, 31 May 1959, *Anderson* 1239!
DISTR. U 3; K 1, 4, 6, 7; T 1–3, 5–8; Z; also central and southern Africa
HAB. Still or slow-flowing freshwater; 0–1700 m.

SYN. *Udora cordofana* Hochst. in Sched. Pl. Exs. Kotschyi It. Nubic., No. 170 (1841)
   Lagarosiphon nyassae Ridley in J.L.S. 22: 234 (1886); Guerke in P.O.A. C: 95 (1895); T. Durand & Schinz in Consp. Fl. Afr. 5: 2 (1895); Wright in F.T.A. 7: 4 (1897). Type: Malawi, Lake Malawi, *Laws* 107 (BM, holo.!, BM, K, iso.!)
   L. tenuis Rendle in J.L.S. 30: 380, t. 31/1–7 (1895); Wright in F.T.A. 7: 3 (1897); U.K.W.F.: 649 (1974). Type: Kenya, Machakos District, E. Ngulia, [Ongalea], Kenani, *Gregory* (BM, holo.!)
   L. crispus Rendle in J.L.S. 30: 381, t. 31/8–17 (1895); Wright in F.T.A. 7: 4 (1897); Chiov. in Fl. Somala 1: 314 (1929); Oberm. in Bothalia 8: 143, fig. 1, 2 (1964) & in F.S.A. 1: 106, fig. 32, 33 (1966). Type: Tanzania, 'Zanzibar' coast–Uyui, *W.E. Taylor* (BM, holo.!)
   L. fischeri Guerke in P.O.A. C: 95 (1895); Wright in F.T.A. 7: 4 (1897). Type: Tanzania, Masai plains, [Massaisteppe], *Fischer* 116 (B, holo.†)
   L. tsotsorogensis Bremek. & Oberm. in Ann. Transv. Mus. 16: 401 (1935). Type: Botswana, Tsotsoroga Pan, *Van Son* 28853 (PRE, holo.)

2. L. muscoides *Harvey* in Hook., Journ. Bot. 4: 230, t. 22 (1841); Wright in F.T.A. 7: 3 (1897); Oberm. in Bothalia 8: 140, fig. 1 (1964); Symoens & Triest in B.J.B.B. 53: 463, fig. 8 (1983). Type: South Africa, Albany, *Zeyher* (TCD, lecto.!)

Stems filiform or terete, 0.2–1.6 mm. in diameter, whitish to dark green. Leaves spirally arranged to subopposite or in whorls, mostly linear, rarely linear-lanceolate, shortly acuminate to narrowly acute, mid-green to dark green, patent, spreading or somewhat recurved, flaccid, somewhat translucent, 4.8–20 mm. long, 0.5–1.2(–1.6) mm. wide, with 2 narrow bands of sclerenchyma fibres on either side of midrib and up to 6 bands of submarginal fibres; margins ± hyaline, denticulate, the teeth usually patent, rarely curved towards the leaf-apex, without a distinct basal portion. Nodal scales ovate to suborbicular, the margins entire or papillose. Male spathe ovate to suborbicular, 3–4 mm. long, 1.7–2.8 mm. wide, pinkish, containing up to 40 flowers. Male flowers: sepals oblong-ovate, 1–1.3 mm. long, ± 0.5 mm. wide, white or pink, translucent; petals ovate, 1–1.7 mm. long, 0.3–0.5 mm. wide, pink or white, translucent; stamens 1–1.7 mm. long. Female spathe narrowly ovate to ovate, 3–4.5 mm. long, ± 2 mm. wide, pinkish. Female flower: sepals oblong-ovate, 1.2–2 mm. long, 0.8–1.1 mm. wide, white or pink, translucent; petals narrowly ovate to ovate, 1.2–1.5 mm. long, 0.6–0.9 mm. wide, purplish pink, translucent; ovary ± 2 mm. long, with 4–8 ovules; perianth-tube up 50 mm. or more long; stigmas 1.1–1.3 mm. long, purple. Fruit narrowly ovoid, 4–10 mm. long, 1.3–3 mm. wide. Seeds narrowly ellipsoid, 2–2.5 mm. long, ± 0.5 mm. wide, brown.

UGANDA. Karamoja District: between Lomaler and Kakamari, June 1930, *Liebenberg* 213!
TANZANIA. Mbulu District: Mbulumbulu, 15 July 1943, *Greenway* 6799!
DISTR. U 1; T 2; also tropical and southern subtropical Africa
HAB. Still or slow flowing freshwater; 1350–1950 m.

SYN. *Hydrilla dregeana* Presl in Abhandl. Bohm. Ges. Wiss., ser. 5, 3: 542 (1845). Type: South Africa, Port Elizabeth District, *Drège* 2276c (P, holo., BM, K, L, iso.!)
   H. muscoides (Harvey) Planchon in Ann. Sci. Nat., sér. 3, 11: 79 (1849)

Lagarosiphon schweinfurthii Caspary in Bot. Zeit. 28: 88 (1870); T. Durand & Schinz in Consp. Fl. Afr. 5: 2 (1895); Wright in F.T.A. 7: 3 (1897); F.P.S. 3: 226 (1956); Hepper in F.W.T.A., ed. 2, 3: 9 (1968). Type: Sudan, Bongo-Land, Gir, *Schweinfurth* 2158 (G, lecto., K!, P, W, isolecto.)

3. **L. hydrilloides** *Rendle* in J.L.S. 30: 381, t. 32/1–7 (1895); Wright in F.T.A. 7: 4 (1897); Symoens & Triest in B.J.B.B. 53: 474, fig. 12. Type: Kenya, Naivasha District, Kariandusi, *Gregory* (BM, holo.!)

Stems terete, 1–2 mm. in diameter, whitish to dark green. Leaves mostly in whorls, rarely spirally arranged or subopposite near base of stem or linear or linear-lanceolate, acute to subobtuse, mid-green to dark green, patent, spreading or somewhat recurved, flaccid, somewhat translucent, 7–32 mm. long, 1.2–1.8(–2.3) mm. wide, with 2 narrow bands of sclerenchyma fibres on either side of midrib and 2–4 bands of submarginal fibres; margins ± hyaline, denticulate, the teeth usually patent, without a distinct basal portion. Nodal scales triangular-subulate, the margins entire. Male spathe ovate, 3–4.5 mm. long, 1.5–2.5 mm. wide, pinkish, containing up to 40 flowers. Male flowers: sepals and petals ovate, 1–1.2 mm. long, ± 0.6 mm. wide, whitish, translucent; stamens ± 1.35 mm. long; anthers ± 0.5 mm. long, purple; filaments ± 1.3 mm. long, colourless. Female spathe narrowly ovate, 2.5–4.5(–5.5) mm. long, 0.8–2.3 mm. wide, pinkish purple. Female flower: sepals ovate, 1–1.3 mm. long, 0.5–0.8 mm. wide, white, translucent; petals narrowly ovate to ovate, ± 1 mm. long, 0.5 mm. wide, white, translucent; ovary ± 3 mm. long, with 5–7 ovules; perianth-tube up to 30 mm. long; stigmas 1–1.4 mm. long, dark purple. Fruit narrowly ovate, 6.5–7.2 mm. long, 0.85–1.8 mm. wide, purplish. Seeds ± ellipsoid, 2.6–3.2 mm. long, 0.7–1 mm. wide, brown.

KENYA. Trans-Nzoia District: Elgon, 28 Feb. 1931, *Lugard* 534!; Laikipia District: near Rumuruti, Aug. 1975, *Powys* 88! & 30 km. N. of Rumuruti, 8 Nov. 1978, *Hepper & Jaeger* 6684!
DISTR. **K** 3; not known elsewhere
HAB. Still or slow-flowing freshwater; 1650–3500 m.

4. **L. ilicifolius** *Oberm.* in Bothalia 8: 145, fig. 1 (1964) & F.S.A. 1: 108, fig. 32 (1966); Symoens & Triest in B.J.B.B. 53: 483, fig. 16 (1983). Type: Botswana, Lake Ngami, Toteng, *Story* 4727 (PRE, holo.)

Stems terete, 0.8–3 mm. in diameter, pink or dark green. Leaves spirally arranged, subopposite or in whorls, often crowded, linear to lanceolate, acute to subobtuse, mid-green to dark green, often strongly recurved, firm, opaque, 4–13 mm. long, (1.2–)2–3.7 mm. wide, without sclerenchyma fibres; margins ± hyaline, denticulate, the teeth projecting at right-angles to the margin, with a colourless, triangular basal portion. Nodal scales linear, the margins entire. Male spathe ovate to suborbicular, 4–4.2 mm. long, 2–3 mm. wide, pink or lilac, containing up to 30 flowers. Male flowers not seen at maturity. Female spathe narrowly ovate to ovate, 2–4.3 mm. long, 0.8–1.8 mm. wide, pink or lilac. Female flower: sepals and petals ovate, 0.8–1.1 mm. long, 0.4–0.5 mm. wide, pale purple, translucent; ovary ± 2 mm. long, with 6–9 ovules; perianth-tube up to 20 mm. or more long; stigmas ± 1 mm. long, purple. Fruit narrowly ovate to ovate, 3–7 mm. long, 1.2–1.5 mm. wide, purplish. Seeds narrowly ellipsoid, 1.5–2 mm. long, ± 0.5 mm. wide, brown.

UGANDA. Mengo District: Lake Victoria, S. tip of Buvuma I., 7 Apr. 1955, *Greenway & Roberts* 8820! & Lake Victoria, Kibanga, Aug. 1914, *Dummer* 1032!
KENYA. Nyanza Province: Lake Victoria, 1928, *M. Graham* H47-28!
TANZANIA. Bukoba District: Lake Victoria, 20 Apr. 1905, *Cunnington* 51!
DISTR. **U** 4; **K** 5; **T** 1; Zaire, Angola, Zambia, Zimbabwe, Namibia and Botswana
HAB. Still or slow-flowing freshwater; 1100 m.

2. HYDRILLA

Rich. in Mém. Cl. Sci. Math. Phys. Inst. France 1811(2): 9, 61, 69, 75 (1812, published 1814); Wright in F.T.A. 7: 1 (1897); Cook & Lüönd in Aquat. Bot. 13: 485 (1982)

Monoecious, rarely dioecious, submerged, freshwater, annual or perennial herbs. Plants often perennating by turions, the turions bulbil-like, arising from tip of stolons or in leaf-axils. Roots simple, adventitious, without root-hairs, arising from nodes. Stems elongate, stoloniferous, or erect, terete, simple or branched from axils; branches 1 per

FIG. 2. *HYDRILLA VERTICILLATA* — 1–3, habit, × ⅔; 4, leaves, × 3; 5, male spathes, × 12; 6, female flowers, × 12; 7, fruit, × 6; 1,5, from *Lye* 3442; 2,4,6,7, from *Aké Assi* s.n.; 3, from *Lind* 177. Drawn by Christine Grey-Wilson.

axil. Leaves sessile, opposite towards base of stem or branch, otherwise in whorls of 3–8(–12), linear to ovate, shortly acuminate to obtuse, patent, spreading or strongly recurved, flaccid, light to dark green, opaque or somewhat translucent, with 1 apical spine; margins hyaline, denticulate, the teeth usually patent; venation consisting of midrib only. Stipules 0. Nodal scales 2 per leaf, narrowly triangular to narrowly lanceolate, fringed with orange-brown hairs. Flowers unisexual. Male spathe solitary in leaf-axil, sessile, appendages at the apex, 1-flowered. Male flowers shortly pedicellate, becoming detached and floating before anthesis; sepals 3, ovate, obtuse, reflexed at anthesis; petals 3, linear to ± spathulate, obtuse, ± reflexed at anthesis; stamens 3, anthers erect, 4-thecous, dehiscing explosively; filaments filiform. Female spathes 1(–2) in leaf-axil, sessile, cylindric but tapering toward apex, 1-flowered. Female flower: sepals 3, free, oblong to ovate, somewhat cucullate, erect; petals 3, free, oblong to ovate, obtuse, erect; staminodes 3, minute, linear or absent; ovary of 3 carpels, 1-locular, placentation parietal; ovules up to 5, anatropous; perianth-tube filiform, carrying perianth to the water surface; styles 3; stigmas 3, linear, usually entire, rarely bifid, papillose. Fruit cylindric, smooth or with lateral spine-like appendages. Seeds up to 5, narrowly ellipsoid, smooth.

A monotypic genus, which shows wide variation in leaf-shape and leaf-size in response to local ecological conditions.

H. verticillata (*L.f.*) *Royle* in Ill. Bot. Himal. 1: 376 (1839); Guerke in P.O.A. C: 95 (1895); F.P.U., ed. 2: 192 (1971); U.K.W.F.: 649 (1974); Cook & Lüond in Aquat. Bot. 13: 485 (1982). Type: India, Linnaean Herbarium no. 1106-1 (LINN, lecto.!)

Stems 1–1.5 mm. in diameter, whitish to dark green. Leaves usually linear to ovate, 2–15 mm. long, 1–3.7 mm. wide. Male spathe 1–1.7 mm. in diameter, whitish, or light brown, somewhat translucent. Male flowers: sepals ovate, 2–3 mm. long, ± 0.6 mm. wide, reddish or brown; petals linear to ± spathulate, ± 2 mm. long, ± 0.2 mm. wide, usually whitish, rarely reddish; stamens up to ± 1.75 mm. long; anthers ± 1 mm. long, whitish, filaments ± 0.75 mm. long. Female spathe up to 6 mm. long, 0.3–0.5 mm. wide, whitish, streaked with purple, somewhat translucent. Female flower: sepals 1.5–2.5 mm. long, 0.3–0.8 mm. wide, white, translucent; petals 1.2–1.5 mm. long, 0.2–0.5 mm. wide, white, translucent; ovary 2–3.5 mm. long; perianth-tube up to 50 mm. or more long; stigmas ± 1 mm. long, whitish to pinkish purple. Fruit 5–15 mm. long, the lateral appendages 1–3 mm. long, dark reddish purple. Seeds 1.9–2.5 mm. long, ± 0.8 mm. wide, brown. Fig. 2.

UGANDA. Kigezi District: Lake Bunyonyi, 31 Dec. 1933, *A.S. Thomas* 1209! & 22 Apr. 1970, *Lye* 5221!; Busoga District: Jinja, 7 Apr. 1955, *Greenway & Roberts* 8820!
KENYA. Nairobi District: cultivated in Nairobi originally from Karen, Jan. 1968, *Cyril*!
TANZANIA. Mwanza District: Mbarika, 15 Aug. 1953, *Tanner* 1630! & Capri Point, 10 Oct. 1953, *Tanner* 1651!
DISTR. U 2–4; K 3, 4; T 1, 4; also Ghana, Ivory Coast, Zaire, Burundi
HAB. Still or slow-flowing freshwater, up to 1.5 m. or more deep; 770–1950 m.

SYN. *Serpicula verticillata* L.f., Suppl. Pl.: 416 (1781)

NOTE. East African material of *H. verticillata* frequently has small, recurved and crowded leaves, and in this state it can easily be confused with *Lagarosiphon ilicifolius*, especially when the two are present in the same habitat. This is exemplified by the many herbarium specimens which consist of mixed collections of the two species. However it is possible to separate them by examining the lower leaves on the stem or branch. In *L. ilicifolius* most of these leaves are spirally arranged, whereas in *H. verticillata* they are in whorls.

3. EGERIA

Planchon in Ann. Sci. Nat., sér. 3,11: 79 (1849); H. St. John in Darwiniana 12: 293 (1961); Cook & Lüönd in Aquat. Bot. 19: 74 (1984)

Dioecious, submerged, freshwater, perennial herbs. Roots simple, adventitious, without root-hairs, arising from nodes. Stems elongate, stoloniferous or erect, terete, simple or branched from axils; branches 1 per axil, subtended at the base by 2, sessile, opposite, scale-like leaves. All other leaves sessile, in whorls of 2–8, linear to lanceolate or narrowly oblong, shortly acuminate to subobtuse, patent, spreading or recurved, flaccid, somewhat translucent, with 1 apical spine; margins green or hyaline, serrulate to denticulate, the teeth usually patent; venation consisting of midrib only. Stipules 0. Nodal scales 2–4 per leaf, ovoid to orbicular with entire margins. Flowers unisexual. Male spathe

FIG. 3. *EGERIA DENSA* — **1**, habit, × ⅔; **2**, male flower, × 3; **3**, female flower, × 2; 1,2 from *Coveny* 4787; 3, from *Ossent* 266. Drawn by Christine Grey-Wilson.

solitary in leaf-axil, sessile, narrowly ovoid, 2–5-flowered. Male flowers long-pedicellate, remaining attached to the plant at anthesis; sepals 3, ovate to suborbicular, obtuse or cucullate, usually reflexed at anthesis; petals 3, widely elliptic to orbicular, rounded, spreading; stamens 9, in whorls of 3, anthers erect, 4-thecous, latrorsely dehiscent; filaments flattened, club-shaped or linear; nectaries 3-lobed, the lateral lobes shorter to longer than the central lobe. Female spathes 1(–2) in leaf-axil, sessile, narrowly ovoid, 1(–2)-flowered. Female flower: sepals 3, ovate to suborbicular, obtuse or cucullate, reflexed; petals 3, widely elliptic to orbicular, rounded, spreading; staminodes 3, linear, papillose. Ovary of 3 carpels, ovoid, 1-locular, placentation parietal; ovules up to 9, orthotropous; perianth-tube filiform, carrying perianth to just above the water-surface; styles 3; stigmas 3, irregularly 2–4-lobed, linear, papillose on adaxial surface, with 2 weakly 3-lobed nectaries at their base. Fruit ovoid, smooth. Seeds ellipsoid, covered in elongated or spine-like cells.

A genus of 2 species, both of which are native to South America.

E. densa *Planchon* in Ann. Sci. Nat., sér. 3, 11: 80 (1849); H. St. John in Darwiniana 12: 297 (1961); U.K.W.F.: 649 (1974); Cook & Lüönd in Aquat. Bot. 19: 80 (1984). Type: Argentina, in ditione Platensi, prope Bonariam, *Tweedie* 10 (K, holo.!)

Stems 0.9–3 mm. in diameter, whitish or green. Scale-like leaves ovate, 1.5–2.5 mm. long, 1.5–2.5 mm. wide, light green or brownish. All other leaves in whorls of 2–5, linear-lanceolate to narrowly oblong, narrowly acute to subobtuse, 1.5–4 cm. long, 0.15–4.8 mm. wide, light to mid-green. Male spathe 7.5–13 mm. long, 1.2–4 mm. wide, greenish white, tinged with pink, somewhat translucent. Male flowers: pedicel up to 80 mm. long; sepals ovate to widely ovate, 2.2–3.5(–5) mm. long, 1.5–2.2(–3) mm. wide, greenish; petals elliptic to orbicular, 6–8(–10) mm. long, 3–8 mm. wide, white; stamens up to 6 mm. long; anthers 0.6–0.7(–1.8) mm. long, yellow; filaments club-shaped, contracted below anthers, papillose above, (0.8–)1.1–1.3(–4.5) mm. long, yellow; nectaries ± 0.8 mm. in diameter, the lateral lobes as long as or longer than the central lobes. Female spathe 9–14 mm. long, 2–4 mm. wide. Female flower: sepals ovate to widely ovate, 3–4 mm. long, 1.6–3 mm. wide, green; petals widely elliptic to suborbicular, 4–8.5 mm. long, 4–7 mm. wide; staminodes 3, linear, papillose, 0.9–2.4 mm. long, yellow to reddish orange; ovary 2–3 mm. long, 1.1–2.2 mm. wide, containing ± 6 ovules; perianth-tube up to 40 mm. or more long; stigmas 2.4–3.8 mm. long, deeply 2-lobed. Fruit 11.5–14.5 mm. long, 4–5.5 mm. wide. Seeds narrowly ellipsoid, 7.2 mm. long, ± 2 mm. wide, brown. Fig. 3.

UGANDA. Mengo District: cultivated in Makerere University Botanic Garden, originally from near Kampala, July 1969, *Lye* 3443!
KENYA. Northern Frontier Province: Marsabit, 14 Feb. 1953, *Gillett* 15108!; Naivasha District: Kinangop, May 1943, *Copley* in *C.M.* 11656!; Nairobi, Nairobi R., 26 Jan. 1961, *Verdcourt* 3044!
DISTR. U 4; K 1, 3, 4; Ghana, South Africa, North America, South America, Europe, Japan and Australasia
HAB. Still or slow-flowing freshwater; 360–2400 m.

SYN. *Elodea densa* (Planchon) Caspary, in Monatsber. Königl. Preuss. Akad. Wiss. Jan. 1857: 48 (1859)

NOTE. *E. densa* is native to southern Brazil, Uruguay and Argentina, but has been widely introduced elsewhere as an exotic plant for garden ponds and aquaria. Although superficially similar to *Lagarosiphon* and *Hydrilla*, it can be distinguished from both these genera by its larger size and large, showy flowers. In Africa only male plants are known at present, and the above description of female flowers is based on that given by Cook & Lüönd in Aquat. Bot. 19: 80 (1984).

4. BLYXA

Thouars, Gen. Nov. Madag.: 4 (1806); Rich. in Mém. Cl. Sci. Math. Phys. Inst. France, 1811(2): 19, 63, 77 (1812, published 1814); Cook & Lüönd in Aquat. Bot. 15: 1 (1983)

Dioecious (perhaps rarely monoecious) or hermaphrodite, submerged, freshwater, annual or perennial herbs. Roots simple, adventitious. Stems corm-like or elongate and rhizomatous or stoloniferous, or erect. Leaves sessile, spirally arranged, radical or cauline, linear to elliptic, acuminate to obtuse, attenuate towards apex, patent, spreading or erect, flaccid, translucent, with up to 28 longitudinal veins; margins green, denticulate, at least towards the apex, the teeth usually patent. Stipules 0. Nodal scales 2 per leaf, narrowly triangular to triangular, with entire margins. Flowers unisexual or bisexual.

Spathe solitary in leaf-axil, sessile or pedunculate, cylindric, 2-lobed at the apex, somewhat translucent, 2–6-veined, with 1–2(–22) flowers. Male flowers long-pedicellate, remaining attached to the plant at anthesis; ♀ and ♂ flowers sessile or subsessile. Sepals 3, linear to linear-lanceolate, acute or cucullate, erect and often forming a tube at anthesis; petals 3, filiform to lanceolate, acute to obtuse, papillose, often erect. Stamens 3 (in ♂ flowers) 6 or 9, in whorls of 3; anthers erect, 4-thecous, latrorsely or introrsely dehiscent; filaments filiform; staminodes (in ♂ flowers only) 3, rudimentary or absent; ovary of 3 carpels, narrowly cylindric, 1-locular, placentation parietal; ovules numerous, anatropous; perianth-tube filiform or narrowly cylindric, carrying perianth to the water surface; styles 3; stigmas 3, simple, linear, papillose on adaxial surface. Fruit elongate, narrowly cylindrical, disintegrating at maturity. Seeds numerous, narrowly ellipsoid, ellipsoid or ovoid, smooth, tuberculate, winged, ribbed or spiny.

A genus of 9 species widely distributed throughout tropical and subtropical regions of the Old World.

Tip of leaf-apex narrowly acute or acuminate; flowers bisexual   *1 B. aubertii*
Tip of leaf-apex obtuse; flowers unisexual . . . . . .   *2. B. hexandra*

1. B. aubertii *Rich.* in Mém. Cl. Sci. Math. Phys. Inst. France, 1811(2): 19, 77, t. 4 (1812, published 1814); Cook & Lüönd in Aquat. Bot. 15: 9 (1983). Type: Madagascar, *du Petit-Thouars* (P, holo.)

Hermaphrodite, perennial. Stems very shortly rhizomatous, or erect and corm-like, simple or forked, up to 1 cm. in diameter. Leaves radical, linear to lanceolate, 2–35(–60) cm. long, 2–8(–12) mm. wide, attenuate to a narrowly acute or acuminate tip, patent or erect, mid-green to dark green, with up to 28 longitudinal veins and faint cross-veins. Flowers bisexual. Spathe: peduncle up to 50 cm. long; body of spathe 4–12 cm. long, 1.5–3 mm. wide, 6-nerved, with 1(–2) flowers. Sepals 5–10 mm. long, ± 1.2 mm. wide, green, often streaked with purple; petals linear, often twisted and folded between the sepals, ± 17 mm. long, ± 0.5 mm. wide, acute, white or reddish; stamens 3, up to 4 mm. long; anthers 1–1.8 mm. long, with a distinct connective-tip up to 1 mm. long; filaments ± 2 mm. long; ovary 40–55 mm. long, ± 1.5 mm. wide, greenish brown; perianth-tube narrowly cylindric, up to 15 cm. or more long; stigmas 10–15 mm. long, white or reddish. Fruit 40–60(–80) mm. long, 1.5–3 mm. wide, brown. Seeds ellipsoid or ovoid, 1.2–2 mm. long, 0.6–0.8 mm. wide, smooth or ribbed wihout spines, or ribbed and spiny with apical and basal spines up to 5 mm. long, light brown.

var. **aubertii**

Seeds ellipsoid or ovoid, 1.2–1.5 mm. long, smooth or ribbed without spines.

TANZANIA.   Rufiji District: Mafia I., 1982, *Kasselman 6/82*
DISTR.   T 6; Mozambique, Madagascar, India, E. and SE. Asia, Japan and Australia
HAB.   Still or flowing water up to 2 m. deep; near sea-level

NOTE.   The above specimen was mentioned in a note added in proof by Cook & Lüönd in Aquat. Bot. 15: 47 (1983). Unfortunately the specimen has not been traced, although its determination is undoubtedly correct. Since no other East African material is currently available, the description here is based on specimens from Mozambique and Madagascar.
    Var. *echinosperma* (C.B. Clarke) Cook & Lüönd, which is found in India, SE. Asia and Australia, differs from the typical variety in having slightly larger and distinctly spiny seeds.

2. B. hexandra *Cook & Lüönd* in Mitt. Bot. Staats., München 16: 485, fig. 1 (1980). Type: Angola, Benguela, Cassipera, *Hess 52/1483* (Z, holo., BM!, Z, ZT, iso.)

Dioecious, usually annual, rarely perennial. Stems corm-like, shortly erect. Leaves radical, linear, 10–37(–40) cm. long, 1–3(–4.5) mm. wide, attenuate to an obtuse tip, patent or erect, light to mid-green; margin denticulate towards apex. Male spathe: peduncle up to 40 cm. long; body of spathe (3.5–)4–5.5(–6.5) cm. long, 2–3.5 mm. wide, 6-nerved, with up to 20 flowers. Male flowers: sepals 4–6(–7) mm. long, 0.7–1.5 mm. wide, acute, green, often streaked with dark purple; petals linear, 1.3–1.5 mm. long, 1–1.3(–1.5) mm. wide, acute or obtuse, white, joined at first, then spreading; stamens 6, up to 3.75 mm. long; anthers 1.2–1.75 mm.long, yellowish, without a connective tip; filaments 1–3 mm. long. Female spathe: peduncle up to 25 mm. long; body of spathe 5.5–8.5 mm. long, 1.8–2.7 mm. wide, 6-nerved, with 1(–2) flowers. Female flowers: sepals 6–10 mm. long, 0.8–1.5 mm.

FIG. 4. *BLYXA HEXANDRA* — **1**, habit of male plant, × ⅔; **2**, male flower, × 4; **3**, female spathe and flower, × 3; **4**, female flower, × 3; **5**, seeds, × 6; 1,2, from *Milne-Redhead & Taylor* 9165; 3–5, from *Gillett* 17421. Drawn by Christine Grey-Wilson.

wide, somewhat cucullate, green, often streaked with dark purple; petals filiform, 2–2.5
mm. long, ± 0.2 mm. wide, white, often withering as stigmas mature; ovary 4–5 mm. long;
perianth-tube filiform or narrowly cylindric, up to 19 mm. long; stigmas up to 27 mm.
long, white. Fruit 4–6 cm. long, ± 2 mm. wide, brown. Seeds narrowly ellipsoid, 1.5–2.5 mm.
long, ± 1 mm. wide, smooth or slightly tuberculate, brown. Fig. 4, p. 11.

TANZANIA.   Songea District: Kwamponjore valley, 14 Mar. 1956, *Milne-Redhead & Taylor* 9165!
DISTR   T 8; Central African Republic, Zaire, Burundi, Zambia and Angola
HAB.   Still water, usually in temporary pools; 1000 m.

NOTE.   *B. hexandra* is closely related to *B. radicans* Ridley. The latter species has not yet been
recorded from the Flora region (cf. Cook & Lüönd in Aquat. Bot. 15: 42 (1983)), although it does
occur close to the Tanzanian border in northern Zambia. It is distinguished from *B. hexandra* by its
stoloniferous, perennial habit and the somewhat cucullate leaf-apex.

5. VALLISNERIA

L., Sp. Pl.: 1015 (1753); Wright in F.T.A. 7: 5 (1897); Lowden in Aquat. Bot. 13: 283 (1982)

Dioecious, submerged, freshwater, perennial herbs. Roots simple. Stems elongate,
shortly stoloniferous or very shortly erect. Leaves sessile, spirally arranged, radical, linear,
obtuse, sheathing at the base, patent or erect, opaque or somewhat translucent, with 3–5
longitudinal veins; margins green, entire to coarsely serrate. Stipules 0. Nodal scales not
seen. Flowers unisexual. Male spathe solitary in leaf-axil, shortly pedunculate, ovoid,
shortly 2-lobed at the apex, translucent, with numerous flowers. Male flowers shortly
pedicellate, breaking free and floating before anthesis; sepals 3, oblong to ovate,
cucullate, reflexed at anthesis; petal 1, minute, linear, rudimentary, erect or reflexed;
stamens 1–3, erect, or divergent, the filaments sometimes united; staminode 1, linear,
rudimentary. Female spathe solitary in leaf-axil, long-pedunculate (the peduncle spirally
contracting after fertilisation), cylindric, 2-lobed at apex, translucent, 1-flowered. Female
flowers: sepals 3, ovate to oblong-ovate, obtuse, erect; petals 3, minute, linear,
rudimentary; staminodes 3, linear, rudimentary or absent; ovary elongate, narrowly
cylindric, 1-locular, placentation parietal; ovules numerous, orthotropous; perianth-tube
absent; styles 3, reduced; stigmas 3, linear, deeply 2-lobed, papillose on adaxial surface.
Fruit elongate, narrowly cylindric, opening by decay of the pericarp. Seeds narrowly
ellipsoid to oblong, striate.

A genus of 2 species, throughout warmer regions of the world.

V. spiralis *L.*, Sp. Pl.: 1015 (1753); Guerke in P.O.A. C: 95 (1895); Wright in F.T.A. 7: 5
(1897); U.K.W.F.: 649 (1974); Lowden in Aquat. Bot. 13: 285, fig. 3 (1982). Type: t. 10/1, 2 in
Michelius, Nov. Pl. Gen. 3: 12 (1729)

Stems creeping and stoloniferous, narrowly terete, 0.6–1.9 mm. in diameter. Leaves
0.5–40(–57) cm. long, 1.7–12 mm. wide, mid-green to dark green with numerous, minute,
longitudinal and transverse reddish brown striations; leaf-margins green, entire to
coarsely serrate. Male spathe with peduncle up to 7 cm. long; spathe 4–5.5 mm. long,
2.3–2.5 mm. wide, light brown, containing up to ± 50 flowers. Male flowers: sepals up to ±
0.5 mm. long. Female spathe: peduncle up to 100 cm. long; body of spathe 8–19 mm. long,
0.95–2 mm. wide, whitish to light brown. Female flower: sepals narrowly ovate to ovate,
1.8–3 mm. long, 0.7–1.8 mm. wide, light brown to reddish brown, often dotted or striated
with dark reddish brown; petals narrowly ovate, ± 0.5 mm. long, ± 0.2 mm. wide; ovary
10–25 mm. long, ± 1 mm. wide; stigmas shallowly or deeply bifid, fringed, ± 2 mm. long.
Fruit 10–30 mm. long, 1–1.5 mm. wide, brown or reddish brown, often dotted or striated
with dark reddish brown. Seeds 1.3–2 mm. long, ± 0.5 mm. wide, papillose, brown, with
reddish brown striations. Fig. 5.

UGANDA.   Busoga District: Lake Victoria, Jinja, Ekunu Bay, 30 Dec. 1955, *Roberts* in E.A.H. 1/1951! &
Jinja, Napoleon Bluff, Apr. 1955, *Greenway* 8828; Mengo District: Entebbe, edge of Lake Victoria,
Jan. 1938, *Chandler* 2091!
KENYA.   Nandi District: Nandi hills, 25 Aug. 1985, *Odero* 1!
TANZANIA.   Lake Province: shores of Lake Victoria, 26 Oct. 1938, *Vanderplank*!; Kigoma District: in
Lake Tanzania, 20 December 1971, *Harris* 6073!; Songea District: Lake Malawi, Lukoma, 1887,
Bellingham!

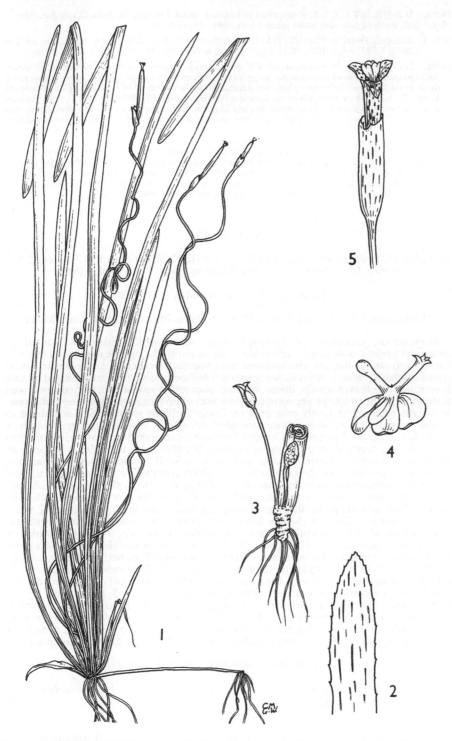

FIG. 5. *VALLISNERIA SPIRALIS* — **1**, habit, × ²⁄₃; **2**, leaf-apex, × 4; **3**, male inflorescence, × 2 ²⁄₃; **4**, male flower, × 18; **5**, female flower, × 4; 1,2, from *Lock* 83/16; 3,4, after Lowden (1982); 5, from *Liebenberg* 733. Drawn by Christine Grey-Wilson.

DISTR.   U 2–4; **K** 3; **T** 1, 4, 7, 8; widespread in tropical Africa, Europe, SE. Asia, Japan, Australia
HAB.   Still, fresh or brackish water up to 3.5 m. deep; 475–1800 m.

SYN.   *V. aethiopica* Fenzl in Flora 27: 311 (1844) nomen & in Sitz. Akad. Wien, Math.-Nat. 51: 139
(1865); F.P.U., ed. 2: 192 (1971). Type: Sudan, *Kotschy* 284 (W, holo., BM!, K!, P, W, iso.)

NOTE.   For many years African material of *V. spiralis* has been referred to as *V. aethiopica*. However
detailed examination of this material suggests that it does not warrant specific recognition, since
there are virtually no differences between African and other material of *V. spiralis*. Two varieties of
V. spiralis have been recognised by Lowden in Aquat. Bot. 13: 287–289 (1982), with African
material being assigned to var. *denseserrulata* Makino. The characters used to distinguish this
variety are the deeply bifid stigmas and the noticeably serrate leaf margin. Unfortunately in
African material this separation seems to be based on only two specimens, and it is virtually
impossible to observe stigma-structure on herbarium samples with any accuracy. The degree of
serration on the leaf-margin is also variable, and although some African material does have
coarsely serrate leaves, many specimens are actually similar to the typical variety, with entire to
denticulate margins. Bearing in mind the range of phenotypic variation exhibited by the family as
a whole, any infraspecific delimition based on herbarium material should be treated with caution.
Further studies need to be carried out using living plants wherever possible.

6. OTTELIA

Pers., Syn. Pl. 1: 400 (1805); Wright in F.T.A. 7: 6 (1897); Hepper in F.W.T.A., ed. 2, 3: 7
(1968); Cook, Symoens & Urmi-König in Aquat. Bot. 18: 263 (1984)

Boottia Wallich, Pl. Asiat. Rariores 1: 51 (1830)

Damasonium Schreber in L.f., Gen. Pl., ed. 8: 242 (1789), nom. illegit., *non* Miller

Monoecious, dioecious or hermaphrodite, submerged, freshwater, annual or
perennial herbs. Roots simple, adventitious. Stems corm-like, simple or forked or (not in
East Africa) rarely elongate, rhizomatous and irregularly branched. Leaves distichous to
spirodistichous, radical, often showing marked differentiation into juvenile and mature
leaves; juvenile leaves sessile, linear to ovate, obovate or spathulate, sometimes armed
with spines or thorn-like projections; mature leaves often petiolate, sheathing at the base;
leaf-blade submerged, partly emergent or floating, elliptic to orbicular, acuminate to
obtuse, or rounded, cuneate, obtuse, rounded, cordate or involute at base, smooth or
scabrid, with up to 15 prominent longitudinal veins connected by cross-veins; margins
green, entire, denticulate, undulate or crispate. Stipules 0. Nodal scales 2–10 per leaf.
Flowers unisexual or bisexual. Spathe solitary in leaf-axil, membranous or coriaceous,
subsessile or pedunculate, the peduncle often spirally contracting after anthesis in ♀ and
♂ flowers, submergent to emergent, often inflated, narrowly cylindric to ovoid, with
wings, ribs, spines, warts or thorns, rarely smooth, 2–6-lobed at the apex, with 1–40
flowers. Male flowers pedicellate, remaining attached to the plant at anthesis; ♀ and ♂
flowers sessile or subsessile. Sepals 3, narrowly lanceolate to ovate, acute to rounded,
often persistent in fruit; petals 3, ovate to orbicular or obovate, truncate, rounded or
obcordate, usually clawed at base; stamens 3–15 or more, in whorls of 3, anthers erect,
4-?theecous, latrorsely dehiscent; filaments often somewhat flattened and fleshy, smooth,
papillose or ciliate; staminodes (0–)3 or more; ovary of 3–20 or more carpels, narrowly
cylindric, ellipsoid or ovoid, 1-locular, placentation parietal; ovules numerous,
anatropous; perianth-tube narrowly cylindric or cylindric; styles 3–20 or more; stigmas
6–40 or more, 2 per style, linear, papillose; nectaries ± 3, located at the base of the styles,
lobe-like, ± 1 mm. long. Fruit fleshy, cylindrical to ovoid, opening by decay of the pericarp
or by regular dehiscence. Seeds numerous, narrowly cylindric to ellipsoid, sometimes
with a short apical projection; testa membranous, densely covered with unicellular hairs.

21 species, occurring widely throughout the warmer regions of the world.

1. Spathe with 3 or more wings . . . .   *1. O. alismoides*
   Spathe with 0–2 wings 2
2. Flowers bisexual 3
   Flowers unisexual 5
3. Mature leaves floating; leaf-base rounded . . . . .   *5. O. somalensis*
   Mature leaves submerged; leaf-base tapering into
   petiole 4

4. Spathe smooth . . . .   *2. O. ulvifolia*
   Spathe covered with spines or prickles . . . . . .   *6. O. verdickii*
5. Leaves lanceolate or elliptic-lanceolate, rarely elliptic,
   usually more than 3 times as long as wide . . . .   *4. O. scabra*
   Leaves elliptic, elliptic-oblong or elliptic-ovate, up
   to 2.5 times as long as wide 6
6. Leaves obtuse to cordate at base, usually less than 6 cm.
   wide . . . .   *3. O. exserta*
   Leaves cuneate at base, usually over 6 cm. wide . . .   *7. O. fischeri*

1. O. alismoides (*L.*) *Pers.*, Syn. Pl. 1: 400 (1805); Wright in F.T.A. 7: 6 (1897); Cook & Urmi-König in Aquat. Bot. 20: 131, fig. 1, 2 (1984). Type: India, Linnaean Herbarium. No. 703/2 (LINN, holo.!)

Monoecious, dioecious or hermaphrodite, annual or perennial. Stems corm-like, simple or rarely forked. Juvenile leaves linear to obovate or spathulate. Mature leaves petiolate; blade submerged or floating, elliptic to widely ovate, 8–17 cm. long, 5.7–20 cm. wide, acute to obtuse or rounded, cuneate to cordate or somewhat involute at base, smooth, with 2–10 prominent longitudinal veins connected by cross-veins, together with many smaller longitudinal veins; leaf-margins entire or denticulate. Nodal scales 4–10 per leaf, ovoid to conical. Spathe: peduncle 10–50 cm. long, 0.17–0.5 cm. wide; body of spathe submerged, sometimes slightly inflated, cylindric to ellipsoid or urceolate, 2–6-lobed at the apex, with 3–12 crisped or wavy wings or ribs. Male flowers several per spathe, with pedicels up to 7 cm. long; ♀ and ♂ flowers sessile. Sepals narrowly triangular to ovate, 0.6–2.4 cm. long, 1.8–5.7(–9) mm. wide, subobtuse to rounded, green, the margin often hyaline; petals ovate to orbicular, 1.8–3 cm. long, 0.5–1 cm. wide, rounded or obcordate, white, white with yellow base, yellow, pink to purple; stamens 3–12; anthers 2.5–4.5 mm. long, yellowish; filaments up to 6 mm. long; staminodes filiform or 2-lobed, up to 4.8 cm. long; ovary of 3–10 carpels, narrowly ellipsoid, 0.7–2.2 cm. long, ± 0.7 cm. wide; perianth-tube 0.6–0.8 cm. long, styles 3–10; stigmas 6–20, up to 6 mm. long, yellowish. Fruit ellipsoid to ovoid, rarely cylindric, 1.5–4(–5) cm. long, 1–2 cm. wide, opening by decay of the pericarp. Seeds narrowly cylindric, 0.9–1.3 mm. long, 0.3–0.5 mm. wide, dark purple to black, with longitudinal striations.

TANZANIA. Masai District: Manyara Ranch, farm dam, 16 Feb. 1965, *Leippert* 5557!; Mbulu District: Tarangire, 8 May 1962, *Polhill & Paulo* 2389!
DISTR. T 2; also Egypt and Sudan; widespread throughout India, E. and SE. Asia and northern Australia as well as scattered localities in Europe, W. Asia and North America
HAB. Still or moving water up to 1 m. deep; 1050–1110 m.

SYN. *Stratiotes alismoides* L., Sp. Pl.: 535 (1753)
   Damasonium alismoides (L.) R. Br., Prodr. Fl. Nov. Holl.: 344 (1810)

NOTE. *O. alismoides* is an extremely variable species, which has caused much taxonomic confusion. However it may be separated from all the other African species by the presence of 3 or more wings on the spathe. Material with unisexual flowers seems to be confined to Asia.

2. O. ulvifolia (*Planchon*) *Walp.* in Ann. Bot. Syst. 3: 510 (1852); Hepper in F.W.T.A., ed 2,3: 7, t. 16 (1968); F.P.U., ed. 2: 192 (1971); U.K.W.F.: 649 (1974); Symoens in Fl. Cameroun 26: 35, t. 9 (1984); Blundell, Wild Fl. E. Afr.: 410, t. 244 (1987). Type: Madagascar, *Lyall* 149 (K, lecto.!)

Hermaphrodite, annual, rarely perennial. Stems erect, corm-like, simple or rarely forked. Juvenile leaves not seen. Mature leaves petiolate; blade submerged or floating, linear-elliptic to ovate-elliptic, 8–45 cm. long, 0.9–11.5 cm. wide, acute to obtuse, cuneate at base, smooth, light to mid-green, with 5–13 prominent longitudinal veins connected by cross-veins, together with many smaller longitudinal veins; margins entire; petiole 6.2–100 cm. long, 0.2–0.8 cm. wide. Nodal scales not seen. Spathe: peduncle 4.2–60 cm. long, 0.3–0.7 cm. wide; body of spathe submerged, not or slightly inflated, linear-oblong, elliptic, ovate or urceolate, 2–5.6(–6) cm. long, 0.4–2.7(–2.8) cm. wide, 2–6-lobed at the apex, with 0–2 wings and several longitudinal ribs or veins, green, somewhat translucent. Flowers 1 per spathe. Sepals linear to linear–lanceolate or linear–oblong, 0.65–1.37(–2) cm. long, 1.6–5.3 mm. wide, acute to obtuse, whitish or green, often with brown longitudinal striations, the margins hyaline; petals obovate, 1.2–2.3(–3) cm. long, 0.8–1 cm. wide, rounded or somewhat truncate, white, white with yellow base, or yellow; stamens 3–6; anthers 1.5–2.3 (–3) mm. long, yellow; filaments up to 6.5 mm. long; ovary of

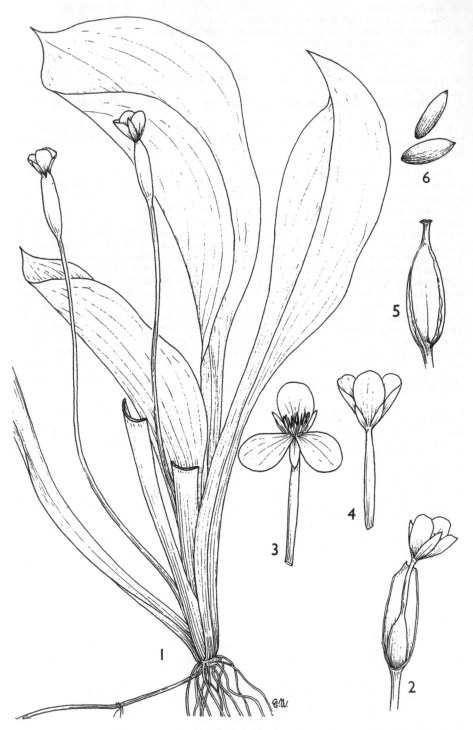

FIG. 6. *OTTELIA ULVIFOLIA* — **1**, habit, × ²/₅; **2**, flower with part of spathe removed, × 1; **3,4**, flower, × 1; **5**, fruit, × ²/₅; **6**, seeds, × 10; 1, from *Tweedie* 3688; 2–4, from *Milne-Redhead & Taylor* 9297; 5,6, from *Schweinfurth* 1159. Drawn by Christine Grey-Wilson.

3–6 carpels, narrowly ellipsoid, 1–1.8 cm. long 0.3–0.4 cm. wide; perianth-tube up to 0.7 cm. long; styles 3–6; stigmas 6–12, up to 20 mm. long, yellowish. Fruit narrowly ellipsoid to ovoid or cylindric, 2–4 cm. long, 1–1.3 cm. wide, green, opening by decay of the pericarp. Seeds narrowly cylindric, 2.1–2.5 mm. long, 0.5–0.7 mm. wide, dark brown. Fig. 6.

UGANDA. Mengo District: 5 km. on Kampala–Entebbe road, Apr. 1930, *Liebenberg* 1555! & Kiagwe, Namanve swamp, Aug. 1932, *Eggeling* 492!; Masaka District: 4–5 km. N. of Lake Nabugabo, 17 Apr. 1969, *Lye & Morrison* 2556!
KENYA. Trans-Nzoia District: Moi's Bridge, 15 Aug. 1963, *Heriz Smith & Paulo* 899!; Kiambu District: Ruiru, 12 July 1952, *Kirrika* 209! & Thika, 18 Sept. 1968, *Faden* 68/733!
TANZANIA. Bukoba District: Ngono R., Dec. 1931, *Haarer* 2376!; Kigoma District: Nguruka, Aug. 1952, *Lowe*!; Songea District: Mtanda, 25 Mar. 1956, *Milne-Redhead & Taylor* 9297!
DISTR. U 1–4; K 3–6; T 1, 2, 4–8; also widespread throughout most of tropical Africa
HAB. Still or slow-moving water up to 2 m. depth; 10–2700 m.

SYN. *Damasonium ulvifolium* Planchon in Ann. Sci. Nat., sér. 3,11: 81 (1849)
   Ottelia lancifolia A. Rich., Tent. Fl. Abyss. 2: 280, t. 95 (1850); T. Durand & Schinz, Consp. Fl. Afr. 5: 4 (1895); Wright in F.T.A. 7: 7 (1897). Type: Ethiopia, Shire, *Quartin-Dillon* (P, holo.)
   O. vesiculata Ridley in J.L.S. 22: 237 (1886); T. Durand & Schinz in Consp. Fl. Afr. 5: 4 (1895); Wright in F.T.A. 7: 7 (1897). Type: Angola, Huila, *Welwitsch* 6497 (BM, holo.!)
   O. plantaginea Ridley in J.L.S. 22: 238 (1886); T. Durand & Schinz, Consp. Fl. Afr. 5: 4 (1895); Wright in F.T.A. 7: 7 (1897). Type: Angola, Huila, *Welwitsch* 6469 (BM, holo.!)
   O. massaiensis Guerke in Urban & Graebner, Festschr. Asch.: 545 (1904). Type: Tanzania, Masai District [Massailand], *Fischer* (B, holo.)
   Boottia abyssinica Ridley in J.L.S. 22: 239 (1886); T. Durand & Schinz in Consp. Fl. Afr. 5: 4 (1895); Wright in F.T.A. 7: (1897). Type: Ethiopia, *Schimper* 1452 (BM, holo.!)
   Ottelia abyssinica (Ridley) Guerke in Urban & Graebner, Festschr. Asch.: 540 (1904)

NOTE. *O. ulvifolia* is the commonest species in East Africa, and shows a wide range of variation in leaf-shape, leaf-size and spathe-structure. According to Symoens in Flore du Cameroun 26: 36 (1984), the leaves have been used for a variety of medicinal purposes, including the relief of palpitations and for children with colds.

3. **O. exserta** (*Ridley*) *Dandy* in J.B. 72: 137 (1934). Types: Mozambique, R. Shire, *Kirk*, & Shupanga, *Kirk* (both K, syn.!)

Dioecious, perennial. Stems erect, corm-like, simple or rarely forked. Juvenile leaves somewhat translucent; petiole to 7 cm. long. Mature leaves petiolate; blade submerged, linear-elliptic to ovate, 4.4–15 cm. long, 1.8–4(–6.5) cm. wide, obtuse, cuneate or cordate at base, mid-green, with 6–10 prominent longitudinal veins connected by numerous cross-veins, together with many smaller longitudinal veins; margins entire; petiole 9–20 cm. long, 0.1–1.6 cm. wide. Nodal scales not seen. Male spathe: peduncle 6–30 cm. long, 0.3–1.7 cm. wide; body of spathe submerged, usually slightly inflated, oblong, elliptic to ovate-lanceolate, 4.6–9 cm. long, 0.6–3.9 cm. wide, 2–6-lobed at the apex, with 0–2 wings and several longitudinal ribs or veins, green or purplish, often minutely tuberculate. Male flowers up to 20 or more per spathe; pedicels up to 9.5 cm. long; sepals linear-lanceolate to linear-oblong, 1–2.6 cm. long, 2–6.5 mm. wide, subacute to obtuse, whitish or green, often with brown longitudinal striations, the margins hyaline; petals obovate, 1.9–2.7 cm. long, 1.2–1.3 cm. wide, rounded or somewhat truncate, white with a yellow base; stamens up to 15; anthers 3.2–4 mm. long, yellow; filaments 6–15 mm. long. Female spathe pedunculate, the peduncle 3.5–8 cm. long, 0.6–1.5 mm. wide; body of spathe similar to ♂. Female flowers: sepals and petals similar to those of ♂; ovary of up to 15 carpels, narrowly ellipsoid, 3–5.5 cm. long, 1–1.5 cm. wide; perianth-tube up to 1.5 cm. long, styles up to 15; stigmas up to 30, 9.5–27 mm. long, yellowish. Fruit ellipsoid, 4–10 cm. long, opening by decay of the pericarp. Seeds narrowly cylindric, 3.4–3.8 cm. long, 1.3–1.5 cm. wide, greenish or brown. Fig. 7, p. 18.

KENYA. Northern Frontier Province: Mkondoni, 5 Apr. 1980, *Gilbert & Kuchar* 5921!; Lamu District: Witu Forest, 5 Mar. 1977, *Hooper & Townsend* 1205! & Mararani, 10 Sept. 1961, *Gillespie* 339! & 340!
TANZANIA. Mbulu District: Tarangire National Park, 12 Mar. 1969, *Richards* 24327! & 13 km. from Tarangire, 29 Nov. 1969, *Richards* 24832!
DISTR. K 1, 7; T 2; Malawi, Zambia, Zimbabwe and Angola
HAB. Still freshwater or rarely wet mud; 50–1500 m.

SYN. *Boottia exserta* Ridley in J.L.S. 22: 240, t. 13 (1886)

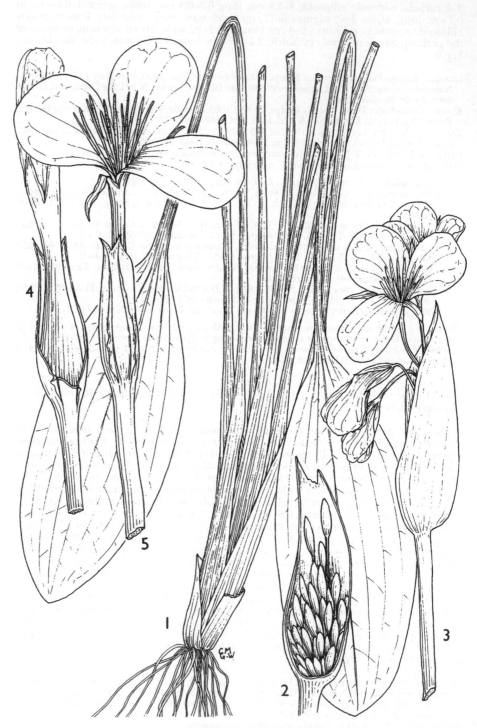

FIG. 7. *OTTELIA FISCHERI* — **1**, habit, × ²⁄₃; **2**, male spathe dissected, × ²⁄₃; **3**, male spathe and flowers, × ²⁄₃; **4**, female spathe dissected, × ²⁄₃; **5**, female spathe and flower, × ²⁄₃; 1, from *Greenway & Polhill* 1154; 2,4, from *Richards* 7526; 3, from *Mahinde* HSM262; 5, from *Burtt* 3620. Drawn by Christine Grey-Wilson.

4. O. scabra *Baker* in Trans. Linn. Soc., Bot. 29: 151 (1875). Type: Sudan, *Grant* 770 (K, holo.!)

Dioecious, annual or rarely perennial. Stems erect, corm-like, simple or rarely forked. Juvenile leaves not seen; mature leaves petiolate; blade submerged, lanceolate, elliptic-lanceolate, rarely elliptic, 13.5–64 cm. long, 4.2–11 cm. wide, obtuse, cuneate or obtuse at base, mid-green to dark green, with 5–8 prominent, longitudinal, often scabrid veins connected by cross-veins, together with many smaller longitudinal veins; margins entire or scabrid; petiole 29–113 cm. long, 0.7–1.3 cm. wide. Nodal scales not seen. Flowers unisexual. Male flowers not seen. Female spathe: peduncle 27–54 cm. long, 0.5–1.2 cm. wide; body of spathe submerged, not or slightly inflated, oblong, elliptic-oblong, lanceolate or ovate, 4.3–7.2 cm. long, 1.3–2.2 cm. wide, 2–6-lobed at the apex, 0–2-winged and with several longitudinal ribs or veins, green, the ribs purple. Female flowers 1 per spathe; sepals linear-lanceolate or lanceolate, 1.8–2.5 cm. long, 3–4.7 mm. wide, acute to obtuse, green, the margins often hyaline; petals obovate, 2.2–3.8 cm. long, 1.7–2 cm. wide, obtuse, white; ovary of 6(–7) carpels, narrowly ellipsoid, 2.5–4 cm. long, 0.4–1 cm. wide; perianth-tube 1.5–3 cm. long; styles 6(–7); stigmas 12–14, 19–20 mm. long, yellowish. Fruit narrowly ellipsoid to ovoid, or cylindric, up to 5 cm. long, opening by decay of the pericarp, 1.6–2 cm. wide, green, disintegrating at maturity. Seeds narrowly cylindric, 3.8–5.5 mm. long, 0.9–1.5 mm. wide, acuminate, dark brown.

UGANDA. Lango District: Atura, Apr. 1943, *Purseglove* 1559!; Bunyoro District: Victoria Nile above Masindi Port, 7 Feb. 1935, *G. Taylor* 3325! & at junction with Lake Albert, 10 Feb. 1935, *G. Taylor* 3356!
DISTR. U 1, 2; Sudan
HAB. Still freshwater; 900 m.

SYN. *Boottia scabra* (Baker) Benth. & Hook. f., G.P. 3: 454 (1883)

5. O. somalensis *Chiov.*, Fl. Somala 2: 413 (1932). Type: Somalia, Oltregiuba, *Senni* 295 (FT, holo.)

Hermaphrodite, annual or perennial. Stems erect, corm-like, simple. Juvenile leaves petiolate; blade submerged, linear to narrowly elliptic, 8–12 cm. long, 0.8–1 cm. wide, light green, somewhat translucent; petiole up to 8 cm. long; mature leaves petiolate; blade floating, lanceolate to ovate, 3.9–10 cm. long, 1.3–6 cm. wide, broadly acute or obtuse, obtuse, rounded or somewhat cordate at base, mid-green to dark green, with 4–7 prominent longitudinal veins connected by cross-veins, sometimes interspersed with many smaller longitudinal veins; margins entire; petiole 10.2–30.4 cm. long, 0.1–0.5 cm. wide. Nodal scales not seen. Spathe: peduncle 30 cm. long, 0.1–0.6 cm. wide; body of spathe submerged, not or slightly inflated, lanceolate to elliptic, 2.4–6.5 cm. long, 0.5–1.1 cm. wide, 2–4-lobed at the apex, 0–2-winged, often with several longitudinal wings or ribs, green or purplish. Flowers 1 per spathe. Sepals linear to linear-lanceolate, 0.9–1.9 cm. long, 0.5–3 mm. wide, acute, green, the margins sometimes hyaline; petals obovate, up to 2.5 cm. long, ± 1 cm. wide, obtuse, white with a yellow centre; stamens (3–)6; anthers ± 5 mm. long, yellow; filaments up to 5 mm. long; ovary of 6 carpels; perianth-tube 2–2.5 cm. long. Fruit and seeds not seen.

TANZANIA. Uzaramo District: 17 km. WSW. of Dar es Salaam, by Pugu road, 3 Aug. 1977, *Wingfield* 4106!; Zanzibar I.: Tomondo [Mtomondo], 18 Jan. 1936, *Vaughan* 2314!
DISTR. T 6; Z; Somalia
HAB. Still freshwater; 0–34 m.

6. O. verdickii *Guerke* in De Wild. in Ann. Mus. Congo, Bot., sér 4,1: 171 (1903). Type: Zaire, Lac Moero, Sept. 1900, *Verdick* (BR, holo.)

Hermaphrodite, perennial. Stems erect, corm-like, simple. Juvenile leaves not seen; mature leaves petiolate; blade submerged, elliptic-lanceolate to elliptic, 15–40(–90) cm. long, 4.5–7(–10) cm. wide, acute, cuneate at base, mid-green to dark green, translucent, with 5–12 prominent, longitudinal, scabrid veins connected by cross-veins, sometimes with many smaller longitudinal veins; margins entire or scabrid; petiole 6–44 cm. long, 0.3–1.2 cm. wide. Nodal scales not seen. Spathe: peduncle 3–11.5 cm. long, 0.4–1.8 cm. wide, spiny; body of spathe submerged, not or slightly inflated, oblong or oblong-lanceolate, 5.4–11 cm. long, 0.9–1.7 cm. wide, 2–6-lobed at the apex, 0–2-winged, often with several longitudinal wings or ribs, strongly tuberculate or spiny, green or purplish.

Flowers 1 per spathe. Sepals oblong to lanceolate, 1.5–2.5 cm. long, 4.2–6.5 mm. wide, subobtuse to acute, green or pale purple-red, the margins often hyaline; petals obovate, up to 3 cm. long, ± 1 cm. wide, white; stamens 9; anthers ± 10 mm. long, yellow; filaments up to 5 mm. long; ovary of 9 carpels, narrowly ellipsoid, 5.7–6.5 cm. long 1–1.8 cm. wide; perianth-tube 2.5–3.5 cm. long; styles 9; stigmas 18, up to 25 mm. long, yellowish. Fruit up to 5 cm. long. Seeds narrowly cylindric, 3–5.5 mm. long, ± 1 mm. wide, with a terminal appendage 5–7 mm. long, brown.

TANZANIA.   Tabora District: Ugalla R., 2 Oct. 1949, *Bally* 7516! & Nov. 1938, *Lindeman* 724! & 16 Aug. 1939, *Hornby* 1023!
DISTR.   **T** 4; Zaire, Zambia and Angola
HAB.   Still freshwater; 1250–1350 m.

NOTE.   A species easily distinguished from the others by its large elliptic or lanceolate leaves and tuberculate or spiny spathes.

7. O. fischeri (*Guerke*) *Dandy* in J.B. 72: 137 (1934). Type: Tanzania, Shinyanga District, Usiha, *Fischer* 588 (B, holo.†)

Dioecious, annual or rarely perennial. Stems erect, corm-like, simple or shortly rhizomatous. Juvenile leaves not seen; mature leaves petiolate; blade submerged, elliptic, elliptic-oblong or elliptic-ovate, 8.4–16 cm. long, (4.5–)6–9.4 cm. wide, obtuse, or rounded, cuneate at base, mid-green to dark green, with 5–9 prominent, longitudinal, veins connected by ascending cross-veins, together with many smaller longitudinal veins; margins entire; petiole up to 38 cm. long, 0.4–0.8 cm. wide, somewhat winged in the lower half. Nodal scales not seen. Flowers unisexual. Male spathe: peduncle 18–28 cm. long, 0.4–1.3 cm. wide; body of spathe submerged, slightly inflated, oblong to ovate, 7–10 cm. long, 1.7–3.2 cm. wide, 2–4-lobed at the apex, 0–2-winged, green. Male flowers pedicellate, up to 15 per spathe; pedicels up to 12 cm. long; sepals linear-lanceolate or lanceolate, 1.6–2.9 mm. long, ± 4 mm. wide, obtuse to subacute, membranous, whitish; petals broadly obovate, 4–4.8 cm. long, ± 2 cm. wide, broadly obtuse, white or yellowish; stamens up to 10; anthers 4–6 mm. long, yellowish; filaments 0.5–1.6 mm. long, papillose, purplish; staminodes 3, linear, up to 2 cm. long, united towards the base. Female spathe: peduncle 30 cm. long, 0.6–1.2 cm. wide; body of spathe submerged, not or slightly inflated, oblong, 4.5–6 cm. long, 0.8–1.5 cm. wide, 2–4-lobed at the apex, 0–2-winged and with several longitudinal ribs or veins, green, the ribs sometimes purplish. Female flowers 1 per spathe; sepals linear-lanceolate or lanceolate, 2–3.2 cm. long, 4–5 mm. wide, acute to obtuse, rarely cucullate, green, the margins often hyaline; petals broadly obovate to suborbicular, 3.8–6 cm. long, 4–5 cm. wide, broadly obtuse, or rounded, white; ovary not seen; styles up to 10; stigmas up to 20, 15–17 mm. long, yellowish. Fruit and seeds not seen. Fig. 7, p. 18.

UGANDA.   Teso District: near Soroti, 30 Dec. 1959, *Napper* 1508! & Soroti, 18 Sept. 1954, *Lind* 442!
KENYA.   Kavirondo District: near Nzoia R., Namalindi, 19 Feb. 1957, *Whitehead* H220/56–7!
TANZANIA.   Mbulu District: Tarangire R., 28 Sept. 1958, *Mahindi* 262!; Singida District: 48 km. from Issuna on Singida–Manyoni road, 13 Apr. 1964, *Greenway & Polhill* 11544!; Dodoma District: Kazikazi, 17 May 1932, *B.D. Burtt* 3620!
DISTR.   **U** 3; **K** 5; **T** 1, 2, 5, 7; Malawi
HAB.   Swamps and shallow fresh water; 1200–1600 m.

SYN.   *Boottia fischeri* Guerke in P.O.A. C: 95 (1895)
         Ottelia sp. A of U.K.W.F.: 649 (1974)

7. ENHALUS

Rich. in Mém. Cl. Sci. Math. Phys. Inst. France, 1811(2): 64, 71, 74 (1812, published 1814); G.P. 3: 454 (1883); Hartog, Sea Grasses of the World: 214 (1970)

Dioecious, submerged, marine, perennial herbs. Roots coarse, simple. Stem elongate, rhizomatous, rarely branched, with very short internodes. Leaves sessile, usually 2–6, distichously arranged at the end of the rhizome, linear, falcate, rounded or obtuse, distinctly sheathing at the base, opaque or somewhat translucent, with numerous longitudinal veins and septate air channels; margins green, entire or serrulate near the apex in young leaf. Stipules 0. Nodal scales 0. Flowers unisexual. Male spathe solitary in the leaf-axil, pedunculate, with numerous flowers, composed of 2 partly connate bracts, the bracts ovate-lanceolate, obtuse, slightly keeled and distinctly hairy on the keel,

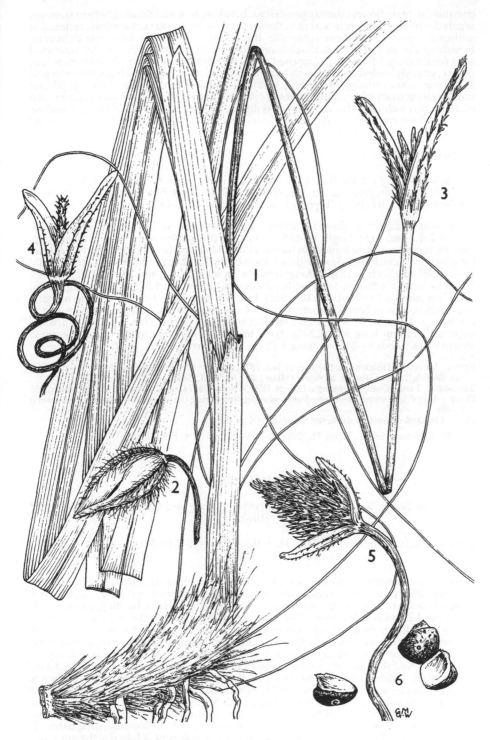

FIG. 8. *ENHALUS ACOROIDES* — **1**, habit, × ²/₅; **2**, male spathe, × ²/₅; **3**, female flower, × ²/₅; **4**, young fruit, × ²/₅; **5**, mature fruit, × ²/₅; **6**, seeds, × 2; 1,3,5, from *Balansa* 3162; 2, from *Parthasanathy* 134; 4, from *Ito* 1454; 6, from *Womersley* NGF 46410. Drawn by Christine Grey-Wilson.

translucent. Male flowers shortly pedicellate, breaking free and floating before anthesis; sepals 3, oblong, acute, reflexed at anthesis; petals 3, ovate, acute, papillose, reflexed at anthesis; stamens 3; anthers subsessile, linear, erect, 1-thecous, latrorsely dehiscent. Female spathe solitary in the leaf-axil, long-pedunculate, the peduncle spirally contracted after fertilisation, 1-flowered, composed of 2, ± free bracts, the bracts oblong-lanceolate, obtuse, strongly keeled and distinctly hairy on the keel, translucent. Female flower: sepals 3, oblong, acute, strongly recurved; petals 3, linear-oblong, acute, erect, papillose; staminodes 0; ovary of 6 carpels, narrowly cylindric, densely hairy, 1-locular, placentation parietal; perianth-tube absent; styles 6, reduced; stigmas 6, subulate, densely papillose except at the apex. Fruit ovoid to subglobose, acuminate, ribbed, densely hairy, opening by decay of the pericarp. Seeds obconical, angular.

A monotypic genus.

E. acoroides (*L.f.*) *Royle*, Ill. Bot. Himal. 1: 453 (1840); Asch. & Guerke in E. & P. Pf. 2,1: 254 (1889); Isaac in Journ. E. Afr. Nat. Hist. Soc. 27: 45, t. 8 (1968); Hartog, Sea Grasses of the World: 215, t. 60 (1970). Type: Sri Lanka, *König* (BM, holo.!)

Stems up to 1.5 cm. in diameter, mid-brown to dark brown. Leaves 30–110(–150) cm. long, 0.7–1.7 cm. wide, bright to dark green; sheaths compressed, up to 15 cm. long, whitish, translucent and somewhat membranous. Male spathe: peduncle up to 5–10 cm. long; bracts ± 5 cm. long, ± 3 cm. wide, with somewhat inrolled margins, light brown. Male flowers with pedicels 3–12 mm. long; sepals ± 2 mm. long, ± 1 mm. wide, white; petals ± 1.75 mm. long, ± 1.25 mm. wide, white; stamens with anthers 1.5–1.75 mm. long, white. Female spathe: peduncle up to 50 cm. long; bracts 3–6 cm. long, 1–2 cm. wide, the margins not inrolled, brown. Female flower: sepals ± 1 cm. long, 0.5 cm. wide, reddish; petals 4–5 cm. long, 3–4 mm. wide, white with a reddish apex; ovary up to 5 cm. long, ± 5 mm. wide; stigmas 10–12 mm. long. Fruit 2.5–5(–7) cm. long, 2.2–3 cm. wide, green or dark brown to almost black. Seeds 10–15 mm. long, ± 12 cm. wide, brown. Fig. 8, p. 21.

KENYA. Mombasa District: Ras Iwetine, 21 Jan. 1974, *Frazier* 912!; Kilifi District: Malindi, Jan. 1954, *van Someren*!; Lamu District: Mokowe [Mkoe], 22 Oct. 1957, *Greenway & Rawlins* 9377!
TANZANIA. Tanga District: Kimanga, 12 Sept. 1976, *Wingfield* 3674!; Rufiji District: Mafia I., *Isaac* 245!
DISTR. **K** 7; **T** 3, 6; widely distributed around coasts bordering the Indian Ocean and the western Pacific
HAB. On sand or mud in sheltered water up to 4 m. deep
SYN. *Stratiotes acoroides* L.f., Suppl. Pl.: 268 (1781); Willd., Sp. Pl. 4: 820 (1806)

8. THALASSIA

König, Ann. Bot. 2: 96 (1805); G.P. 3: 455 (1883); Hartog, Sea Grasses of the World: 222 (1970)

Schizotheca Solms in Schweinf., Beitr. Fl. Aethiop. 1: 194, 246 (1867); Asch. in Linnaea 35: 159, 177 (1867)

Dioecious, submerged, marine, perennial herbs. Stem elongate, rhizomatous or shortly erect, the latter produced at regular intervals from nodes along the rhizome. Roots 1 per node where erect stems produced from the rhizome, simple, sand-binding, covered with fine hairs. Leaves sessile, usually 2–6, distichously arranged on the erect stems, linear, somewhat falcate, rounded or obtuse, distinctly sheathing at the base, opaque or somewhat translucent, with up to 19 longitudinal veins and numerous fine, longitudinal air channels, the longitudinal veins connected by cross-veins; margins green, entire. Stipules 0. Nodal scales 0. Flowers unisexual. Male spathes 1–2 in the leaf-axil, pedunculate, 1-flowered, composed of 2 bracts connate on one side only, the bracts oblong to lanceolate, acute to obtuse, entire or serrulate, translucent. Male flowers shortly pedicellate, remaining attached to the plant; tepals 3, elliptic, cucullate, strongly recurved at anthesis; stamens 3–12; anthers nearly sessile, oblong, erect, 2–4-thecous, latrorsely dehiscent. Female spathe solitary in the leaf-axil, pedunculate, 1-flowered, composed of 2 connate bracts, cylindric, but tapering at both ends, translucent, 2-lobed at the apex, the lobes acute to obtuse. Female flowers almost sessile; tepals 3, similar to the those of ♂; staminodes 0; ovary of 6–8 carpels, conical, muricate, 1(–3)-locular, placentation parietal;

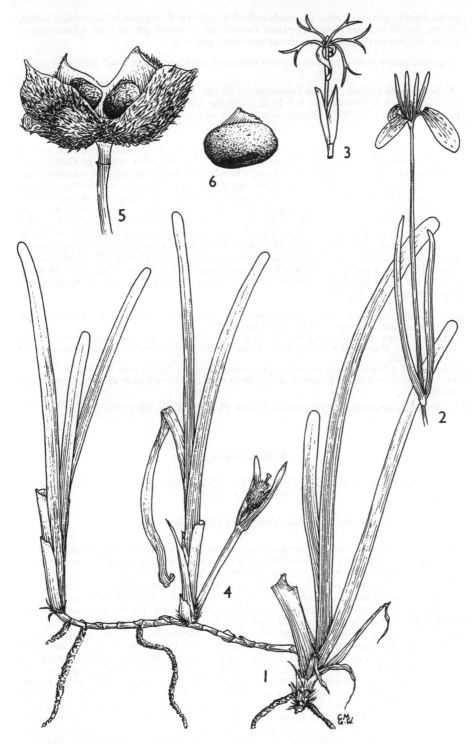

FIG. 9. *THALASSIA HEMPRICHII* — **1**, habit, × ⅔; **2**, male flower, × 2; **3**, female flower, × ⅔; **4**, female spathe and young fruit, × ⅔; **5**, dehisced fruit, × 2; **6**, seed, × 4; 1,4, from *Rajeshwari & Lakshmanan* 3; 2, from *Ostenfeld* 1920; 3, after Isaac (1968); 5,6, from *Burkhill* 1118. Drawn by Christine Grey-Wilson.

ovules several; perianth-tube narrowly cylindric; styles 6–8; stigmas 6–8, deeply 2-lobed, filiform, papillose. Fruit globose, beaked, densely hairy, opening by stellate dehiscence of the carpels. Seeds conical with a thickened basal portion.

2 species, widely distributed in the warmer regions of either the Old or New World.

T. hemprichii (*Solms*) *Asch.* in Petermann's Geogr. Mitth. 17: 242 (1871); Bot. Zeit. 33: 765 (1875); Asch. & Guerke in E. & P. Pf. 2, 1: 254, fig. 188 (1889); Isaac in Journ. E. Afr. Nat. Hist. Soc. 27: 44, t. 7 (1968); Hartog, Sea Grasses of the World: 232, fig. 61 (1970). Type: Ethiopia, Eritrea, Massawa, 1820–26, *Ehrenberg* (B, holo.†, BM!, K!, L!, P, iso.)

Rhizome terete, 2–5 mm. in diameter, greenish to light brown; erect stem terete, up to 5 mm. in diameter. Leaves 4–25(–40) cm. long, 2.5–7(–11) mm. wide, bright to dark green; leaf-sheaths compressed, up to 9 cm. long, whitish, somewhat translucent, membranous and persistent. Male spathe with peduncle up to 3 cm. long; spathe-bracts with margins entire, rarely serrulate near apex, 1.7–2.5 cm. long, ± 5 mm. wide, light brown striated with dark reddish brown. Male flowers: pedicels 2–3 cm. long; tepals 7–8 mm. long, ± 3 mm. wide, light brown or colourless, dotted or striated with reddish brown; stamens (3–)6–9(–12); anthers 7–11 mm. long, yellow. Female spathe with peduncle up to 1.5 cm. long; bracts 1.7–2.5 cm. long, ± 1 cm. wide, light brown striated with dark reddish-brown. Female flowers: ovary of 6 carpels up to 10 mm. long, perianth-tube 20–30 mm. long; stigmas 10–15 mm. long, light brown, becoming recurved at maturity. Fruit 2–2.5 cm. long, 1.7–3.25 cm. wide, light green. Seeds 3–9, ± 8 mm. long and wide, greenish, the basal portion brown. Fig. 9, p. 23.

Kenya. Kwale District: Diani, 6 Jan. 1974, *Frazier* 871!; Kilifi District: Kikambala, 9 Jan. 1974, *Frazier* 885! Lamu District: Kiunga, 29 Aug. 1961, *Gillespie* 319!
Tanzania. Tanga District: Kigombe, 16 Aug. 1932, *Geilinger*! & coast near Bomalandani [Bomandani], *Drummond & Hemsley* 3688!; Lindi District: about 6.5 km. N. of Lindi, *Milne-Redhead & Taylor* 7480!
Distr. K 7; T 3, 8; widespread around tropical and subtropical coasts in the Old World
Hab. Fine mud to clean coral-sand on reef platforms and sublittoral flats from low water mark to 5 m. depth

Syn. *Schizotheca hemprichii* Solms in Schweinf., Beitr. Fl. Aethiop. 194, 246 (1867)

9. HALOPHILA

Thouars, Gen. Nov. Madag. 2: 2 (1806); G.P. 3: 455 (1883); Hartog, Sea Grasses of the World: 238 (1970)

Lemnopsis Zoll., Syst. Verz. 1: 74 (1854), *non* Zipp. (1829)

Monoecious or dioecious, submerged, marine perennials. Stems elongate, rhizomatous or shortly erect, the latter produced from nodes at regular intervals along the rhizome. Roots 1(–3) at each node where the erect stems produced, simple, sand binding, covered with fine hairs. Leaves in pairs, whorls or distichously arranged on the erect stems, sessile or petiolate; blade linear to ovate, acute, obtuse or rounded, cuneate or rounded at the base, opaque or somewhat translucent, usually glabrous, rarely sparsely pubescent, with 3 longitudinal veins, often connected by ascending cross-veins; margins green, entire or serrulate. Stipules 2, obtuse to suborbicular or transversely elliptic, membranous, subtending each pair or whorl of leaves. Nodal scales 0. Flowers unisexual. Male and ♀ spathes similar, solitary in leaf-axils, sessile, 1–2-flowered, rarely with 1 flower of each sex in the same spathe, composed of 2 free, overlapping bracts, the bracts elliptic to suborbicular, acuminate, obtuse, often keeled, translucent, the margins entire or serrulate. Male flowers pedicellate, remaining attached to the plant; tepals 3, ovate, cucullate; stamens 3, anthers sessile, linear-oblong, erect, 2–4-thecous, latrorsely or extrorsely dehiscent. Female flowers with tepals 3, reduced to lobes at the apex of perianth-tube; staminodes 0; ovary of 3–5 carpels, ellipsoid or ovoid, 1-locular, placentation parietal; ovules few to many; perianth-tube narrowly cylindric; styles 3–5; stigmas 3–5, filiform, entire. Fruit ovoid to globose, membranous, opening by decay of pericarp. Seeds subglobose to globose, tuberculate, reticulate or smooth.

8 species, of which 4 are found on the E. African coast. All the East African species belong to sect. *Halophila*, which is characterised by plants with very short, erect stems, one pair of petiolate leaves at each node and ascending cross-veins on the leaf-lamina.

1. Leaves linear; stipules 8–17 mm. long   . . . . .   *4. H. stipulacea*
   Leaves elliptic to obovate; stipules 0.35–6 mm. long 2
2. Erect stems 1–10 mm. long; leaf-margins serrulate   *3. H. decipiens*
   Erect stems ± absent; leaf-margins entire   3
3. Leaves with 11–15 pairs of cross-veins ascending at an
   angle of 45–60°; . . . .   *1. H. ovalis*
   Leaves with 3–11 pairs of cross-veins ascending at an angle
   of 70–90°   . . . .   *2. H. ovata*

1. H. ovalis (*R. Br.*) *Hook. f.*, Fl. Tasman. 2: 45 (1858); Asch. & Guerke in E. & P. Pf. 2,1: 249, fig. 182 (1889); Isaac in Journ. E. Afr. Nat. Hist. Soc. 27: 42, t. 6A–G (1968); Hartog, Sea Grasses of the World: 240, fig. 62 (1970). Type: Australia, Queensland, *R. Brown* 5816 (K, holo.!, BM, iso.!)

Dioecious. Roots 1(–2) per node. Rhizome narrowly terete, often dichotomously branched, 0.2–1.5 mm. in diameter, whitish or greenish. Erect stems ± absent. Leaves in pairs, 1 pair on each erect stem; blade obovate, ovate, oblong-elliptic or spathulate, rarely linear, 7–16.5(–30) mm. long, 2.5–6.9(–13) mm. wide, obtuse, rounded or subapiculate at apex, cuneate or rounded at base, glabrous, opaque or somewhat translucent, mid-green to dark green, with 2 longitudinal veins running along or very near the margin, connected by 7–25 pairs of cross-veins ascending at an angle of 45–60°; margins entire; petiole 1.5–2 cm. long. Stipules obovate or suborbicular, 3–6 mm. long, 1.5–3 mm. wide, slightly notched, auriculate at base, often distinctly keeled, glabrous, greenish or whitish. Spathe-bracts ovate, 3–4.5(–5) mm. long, 0.8–1.3 mm. wide, acute, often keeled, greenish or whitish, somewhat translucent. Male flowers: pedicel up to 2.5 cm. long; tepals elliptic or ovate, 3–4 mm. long, 2–3 mm. wide, obtuse, spreading or reflexed, yellowish; anthers 2–4 mm. long. Female flowers: ovary of 3 carpels, ovoid, 1–3 mm. long; perianth-tube 3–10 mm. long, the lower portion persisting as beak in fruit; styles 3; stigmas 3, 10–25 mm. long. Fruit globose, 3–4 mm. in diameter, with beak up to 6 mm. long, whitish or brown. Seeds 20–30, globose, ± 1 mm. in diameter, tuberculate or reticulate, light brown. Fig. 10, p. 26.

subsp. **ovalis**

Leaves oblong-elliptic to ovate with 11–15 pairs of cross-veins.

KENYA. Mombasa District: Nyali Beach, 19 Jan. 1962, *Dodd*!; Kilifi District: Watamu, Turtle Bay, 26 Aug. 1965, *Isaac* 66!; Lamu District: Kiunga, 29 Aug. 1961, *Gillespie* 317!
TANZANIA. Zanzibar I., Chwaka [Chakwe] beach, 21 Aug. 1959, *Napper* 1267! & Marahubi, 14 Nov. 1963, *Faulkner* 3306! & Mkokotoni, 13 Mar. 1956, *Oxtoby*!
DISTR. K 7; T 6; Z; widely distributed around coasts in warmer regions of the Old World.
HAB. On fine mud to coarse coral rubble from mid-tidal level to 12 m. depth

SYN. *Caulinia ovalis* R. Br., Prodr. Fl. Nov. Holl. 1: 339 (1810)
     Kernera ovalis Schultes, Syst. Veg. 7: 170 (1829)

NOTE. *H. ovalis* shows a wide range of variation, and has been divided into 4 further subspecies by den Hartog (Sea Grasses of the World: 251 (1970)). 3 of these are confined to either Australia (subsp. *australis* (Doty & Stone) Hartog), the Hawaiian Islands (subsp. *hawaiiana* (Doty & Stone) Hartog) or the western Pacific (subsp. *bullosa* (Setchell) Hartog). Subsp. *linearis* (Hartog) Hartog occurs along the coast of Mozambique and could also be present in East Africa. It is distinguished from the typical subspecies by its linear leaf-blades with ± 10 pairs of ascending cross-veins.

2. H. ovata *Gaudich.* in Freyc. Voy. Monde, Bot., t. 40, fig. 1 (1827); Hartog, Sea Grasses of the World: 251 (1970). Type: Marianne I., *Gaudichaud* (P, holo.!, L, iso.!)

Dioecious. Roots 1 per node. Rhizome filiform or narrowly terete, unbranched, 0.2–0.6(–0.8) mm. in diameter, whitish or greenish. Erect stems hardly developed. Leaves in pairs, 1–2(–3) pairs on each erect stem; blade obovate, ovate or oblong-elliptic, 7–14 mm. long, 3–5 mm. wide, obtuse or rounded, rarely apiculate, obtuse or cuneate at base, glabrous, somewhat translucent, light to mid-green, rarely dark green, with 2 longitudinal veins running along or very near the margin, connected by 3–11 pairs of cross-veins ascending at an angle of 70–90°; margins entire; petiole 0.5–2 cm. long. Stipules

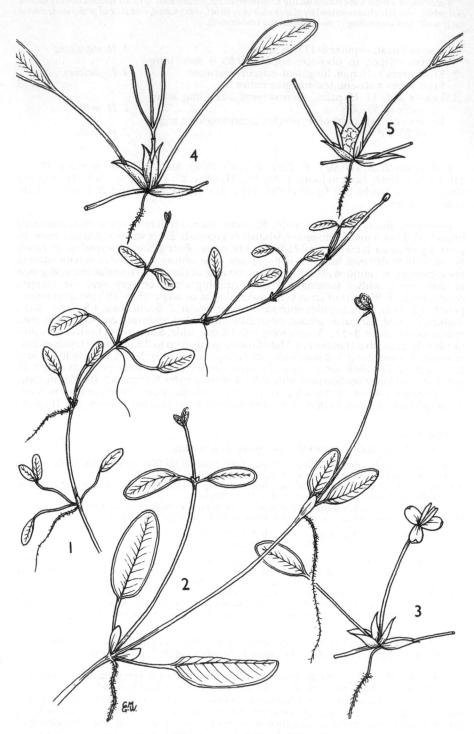

FIG. 10. *HALOPHILA OVALIS* — **1**, habit, × ⅔; **2**, habit, × 2; **3**, male spathe and flower, × 2; **4**, female spathe and flower, × 1; **5**, fruit, × 2; 1,2, from *Faulkner* 3306; 3–5, after Isaac (1968). Drawn by Christine Grey-Wilson.

suborbicular to transversely elliptic, 2–4 mm. long, up to 2 mm. wide, acute or obtuse, slightly auriculate at base, glabrous, greenish or whitish. Spathe-bracts ovate, 2.5–3 mm. long, acute or acuminate, keeled, whitish, translucent. Male flowers: pedicel up to 2 cm. long; tepals elliptic or ovate ± 3mm. long, ± 1 mm. wide, obtuse, spreading; anthers ± 2 mm. long. Female flowers: ovary of 3 carpels, ovoid, 1.5–3 mm. long, ± 2 mm. wide; perianth-tube 2–5 mm. long, the lower portion persisting as beak in fruit; styles 3; stigmas 3, 6–15 mm. long. Fruit ellipsoid to globose, 2–4 mm. long with beak up to 6 mm. long, whitish or brown. Seeds up to 20, globose, 0.5 mm. in diameter, tuberculate or reticulate, yellowish brown.

KENYA. Kwale District: Gazi, 11 Dec. 1965, *Isaac* 116!; Lamu District: Lamu, 1 July 1965, *Isaac* 21! & Mokowe, 30 June 1965, *Isaac* 100!
DISTR. **K** 7; widely distributed around SE. Asian and Australian coasts
HAB. On sand and mud in sheltered sea-water up to 2 m. depth.

SYN. *Lemnopsis minor* Zoll., Syst. Verz. 1: 75 (1854). Type: Indonesia, Lesser Sunda I., Flores, *Zollinger* 3334 (P, holo., BM!, L!, U, iso.)
  Halophila ovalis (R. Br.) Hook. f. var. *ovata* Balfour in Trans. Bot. Soc. Edinb. 13: 335 (1879)
  H. minor (Zoll.) Hartog, in Fl. Males., ser. I, 5: 410. fig. 17b (1957) & in Pac. Pl. Area. 2: 214 (1966); Isaac in Journ. E. Afr. Nat. Hist. Soc. 27: 42, t. 6H–P (1968)

3. H. decipiens *Ostenf.* in Bot. Tidsskr. 24: 260, fig. (1902); Asch. in Neum. Anl. wiss. Beob. Reisen, ed. 3, 2: 395 (1906); Hartog, Sea Grasses of the World: 254 (1970). Type: Thailand, off Koh Kahdat, *Schmidt* 540 (C, holo.!, L, iso.!)

Monoecious. Roots 1 per node. Rhizome narrowly terete, unbranched, 0.1–0.4 mm. in diameter, whitish or greenish. Erect stems 1–10 mm. long. Leaves in pairs, 1 pair on each erect stem; blade elliptic to oblong-elliptic, 10–21 mm. long, 4.5–8 mm. wide, obtuse or rounded at apex, cuneate at base, often sparsely pubescent, rarely glabrous, somewhat translucent, light to mid-green, rarely dark green, with 2 of the parallel veins often running up to 0.5 mm. below the margin, connected by 6–9 pairs of cross-veins ascending at an angle of 40–70°; margins serrulate; petiole up to 15 mm. long. Stipules obovate to suborbicular, 0.35–4 mm. long, 0.3–2.8 mm. wide, obtuse or rounded, keeled, sparsely pubescent on keel, greenish or whitish. Spathe-bracts ovate, 2.5–4.1 mm. long, ± 1.5 mm. wide, acuminate, keeled, whitish, translucent, containing 1 ♂ and 1 ♀ flower. Male flowers: pedicel up to 3 mm. long; tepals elliptic to ovate, 0.3–1 mm. long, ± 0.5 mm. wide, obtuse, spreading, whitish; anthers ± 1 mm. long; filaments up to 4 mm. long. Female flowers: ovary of 3 carpels, ovoid, 1 mm. long, perianth-tube 1–2 mm. long, the lower portion persisting as beak in fruit; styles 3; stigmas 3, ± 2.5 mm. long. Fruit ellipsoid to subglobose, ± 2 mm. long, ± 1.8 mm. wide, with beak up to 2 mm. long, whitish. Seeds up to 30, ovoid, 0.2 mm. long, light green or light brown.

TANZANIA. Zanzibar I., Mkokotoni, 8 Sept. 1955, *Oxtoby* 908!
DISTR. **Z**; widely distributed around tropical and subtropical coasts bordering the Indian and Pacific Oceans, together with the Caribbean.
HAB. Soft mud and fine coral sand in sea-water up to 85 m. depth

SYN. *Halophila decipiens* Ostenf. var. *pubescens* Hartog, Fl. Males., ser. I, 5: 411 (1957) & Acta Bot. Neerl. 8: 485 (1959) & Pac. Pl. Area 2: 216 (1966)

4. H. stipulacea (*Forsskål*) *Asch.* in Sitz. Ges. Nat. Freunde Berlin 1867: 3 (1867); E. & P. Pf. 2, 1: 249, fig. 183 (1889); Hartog, Sea Grasses of the World: 258, fig. 63 (1970). Type: Yemen, *Forrskål*, not located

Dioecious. Roots 1 per node. Rhizome narrowly terete, unbranched, 0.2–2 mm. in diameter, whitish or greenish. Erect stems narrowly terete, up to 1.5 mm. long, 0.2–0.5 mm. wide. Leaves in pairs, 1 pair on each erect stem; blade linear to oblong or linear-elliptic, 1.5–8 cm. long, 2–6 mm. wide, often slightly falcate, obtuse or rounded at apex, cuneate at base, glabrous, somewhat translucent, light to mid-green, rarely dark green, with 2 of the longitudinal veins running just below the margin, connected by numerous pairs of somewhat inconspicuous cross-veins ascending at an angle of 40–60°; margins green, serrate at least near the apex; petiole 0.4–1.5 cm. long. Stipules obovate, 8–17 mm. long, 4.2–10 mm. wide, notched, strongly keeled, glabrous, greenish or whitish. Male flowers not seen. Female flowers: spathe-bracts ovate, ± 10 mm. long, ± 0.7 mm. wide, emarginate, keeled, the keel denticulate, whitish, translucent; ovary of 3 carpels, ovate or ovate-elliptic, 2–3 mm. long, perianth-tube up to 6 mm. long, the lower portion persisting as beak in fruit; styles 3; stigmas 3, 20–25 mm. long. Fruit ellipsoid, ± 5 mm. long, ± 2 mm. wide, with beak up to 6 mm. long, whitish or brown. Seeds up to 40, subglobose, 0.5–0.6 mm. long, ± 0.5 mm. wide, reticulate, light brown.

KENYA. Kwale District: Gazi, 18 Aug. 1965, *Isaac* 61; Mombasa District: 9.5 km. N. of Mombasa, Nyali Beach, 18 Apr. 1950, *Rayner* 288!; Lamu District: Oseni [Osine], 9 Oct. 1957, *Greenway & Rawlins* 9325!

DISTR. K 7; Egypt, Sudan, Ethiopia, Madagascar, Mauritius, Rodriguez, Saudi Arabia, Bahrain and India

HAB. On sand and mud in sheltered sea-water up to ± 7 m. deep

SYN. *Zostera stipulacea* Forsskål, Fl. Aegypt.-Arab.: 158 (1775); Vahl, Enum. Pl. 1: 15 (1804); Willd., Sp. Pl. 4: 179 (1805); Pers., Syn. Pl. 2: 529 (1807)
     Halophila balfourii Soler., Beih. Bot. Centralbl. 30,1: 47 (1913); Isaac in Journ. E. Afr. Nat. Hist. Soc. 27: 42, t. 6Q–T (1968). Type: Rodrigues, *Balfour* (K, holo.!, BM!, P, iso.)
     Thalassia stipulacea König in Ann. Bot. 2: 97 (1805); Sprengel, Syst. Veg., ed. 16, 2: 271 (1825); Kunth, Enum. Pl. 3: 120 (1841); Miq., Fl. Ind. Bat. 3: 226 (1855)

NOTE. This the most distinct of the four E. African species of *Halophila*, with narrower leaf-blades and longer, more conspicuous stipules. East African material of *H. stipulacea* has often been treated as a separate species, *H. balfourii* to distinguish it from plants found in the Red Sea and Persian Gulf. The main difference between the two is the presence of stiffer leaves in the latter, which often become bullate with age. It is unlikely that such a difference warrants the recognition of two separate species, although they may represent different subspecies within *H. stipulacea*. Further collections are needed before any division can be made with certainty.

INDEX TO HYDROCHARITACEAE

T - #0688 - 101024 - C0 - 244/170/2 - PB - 9789061913498 - Gloss Lamination